Published by MathsGroundWork

http://www.mathsgroundwork.co.uk

© PD Burnett 2008 and 2019

This book is copyright.

First published 2008

Revised 2019

Printed in the United Kingdom

Typeface Calibri 11pt

The publisher has used its best endeavours to ensure that URL's for external websites referred to in this book are correct and active at the time of going to press. However, we have no responsibility for the websites and can make no guarantee that a site will remain active or that the content will remain appropriate. The experimental procedures outlined in this book should only be undertaken if the teacher responsible has deemed it safe to do so.

Thanks to Helen Booth, Dawn Pirie, Gordon Doig, Rhona McKinnon and Seonaid Dey for invaluable help with the illustrations. The photograph of the LHC tunnel on the inside back cover is courtesy of CERN.

An important reminder:

This book was originally written for the previous Advanced Higher Physics Investigations course (up to about roughly 2015). Newton's Laws remain unchanged, but the Report writing section has been adjusted to be a better fit to the new requirements. Make sure you refer to the new syllabus and understand what is required. The suggestions for projects, experimental tips and the more social aspects of the Investigation, should remain helpful.

Good luck.

Contents

Chapter 1   The Big Picture   4

Chapter 2   Choosing a Project   6

Chapter 3   The Daybook   59

Chapter 4   Writing the Report   67

    Section 1   The Introduction   71

    Section 2   Procedures   75

    Section 3   Results   78

    Section 4   Conclusions   81

    Section 5   Presentation   85

Chapter 5   In the Lab   87

    Section 1   Safety   87

    Section 2   Home from home; the Physics Lab   89

    Section 3   Do you believe in Hobgoblins?   91

    Section 4   Instruments   92

Chapter 6   Handling Uncertainty   107

Chapter 7   Drawing Graphs   118

Chapter 8   What if it goes Wrong?   122

Organising your Time

Opinions vary amongst teachers when to do the Investigation. Some like to get the Theory out of the way first and spend the rest of the time on the Investigation. The downside of this is that you cover the hard parts of the Theory Units before you're experienced enough to understand it. Others like to run them in parallel. In this case, the Investigation is stretched out over a longer period and can lose focus (you might also have an Advanced Higher Chemistry Investigation to do, and it's easy to put off something when it's still 'early days'). The best choice is the one the suits that particular group of students in that year. The teacher in charge should be flexible in their approach, rather than simply repeating the previous well worn regime.

In practice, it's common for the Investigation to end in a rush. Here's the sorry tale;

- student thinks experiments will only take a few double periods
- experiments don't work as expected
- can't get exclusive use of an important piece of equipment because other teachers are demonstrating it to another class
- the 'daybook' remains stubbornly sparse
- University 'Open Days' coincide with Physics periods
- days are lost due to snow closures
- it gets near the end of March and the experiments aren't complete
- teacher reminds students that he/she needs time to look over the write-up before the Easter holiday so they can make suggestions for improvements
- student promises to spend Easter holidays on the write-up, but social diversions get in the way
- first draft is handed in on the Monday of the week due for posting to the SQA, on the assumption that the teacher can go without sleep for 72hrs.
- frantic 72hrs of e-mails with attachments, going back and forth between teacher and student(s)
- suddenly remember the sparse 'daybook'

It doesn't have to be like that, but every teacher will recognise the sequence. A wiser alternative is:

- definite project chosen and researched before Christmas
- experiments completed by end of February
- daybook kept up to date every session
- write-up during March
- **draft handed in to teacher** at least one week before Easter

- final corrections made during Easter holidays
- report submitted in final form on Monday morning of last week

Teachers are reluctant to treat sixth year students in the same way as junior pupils. Demonstrate a responsible attitude to them and they will give you a lot of slack. A student who shows real interest and enthusiasm for their work will also discover the teacher gives the same in return. This is the basis of a successful project. Others, however, need spoonfed; to such an extent that the SQA encourages teachers to hand out timesheets to students, checking when each stage is reached. This shouldn't be necessary. The role of the teacher should be as a guide rather than a nursemaid. After the pressure of the Higher exams in fifth year, it's tempting to ease off in sixth year; after all, you might already have the firm offer of a place at a University. Don't let that lead to a wasted year and eventual disappointment.

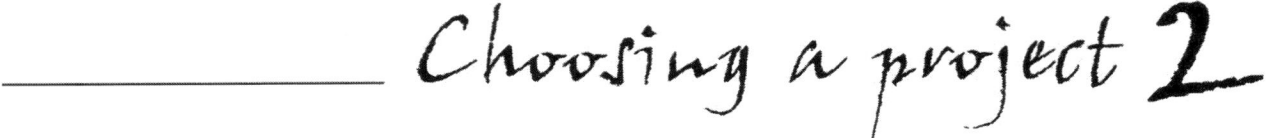

Choosing a project 2

Your teacher will give you the option of coming up with your own idea for an Investigation, picking one from a list, or being assigned one. If you decide not to do an Investigation (!!!) or do not submit one in time (!), then you will not see a course award on your certificate in August (just the units which you have passed). No matter how well you do in the Theory part, this would be a bad career move if your University place depended upon it.

Different projects suit different students, so choose one which suits **you**. Many schools will keep copies of previous years Reports to give you an idea of what's expected. Your teacher will indicate which ones obtained good marks and which ones didn't. Reports from before the millennium will seem thicker on average since more marks were awarded in those days (and students slaved over them even more than you will). Have a quick flick through them to get the 'feel' of what's expected. Notice the difference in presentation; from the crisp business-like document, to the scruffy heap of paper held together with string.

Should you do one big experiment or a series of smaller ones? Until recently, it didn't matter. But now the SQA have declared a preference for 3 or 4 smaller experiments; in their words, grammar and emphasis:

__Minimum of 3 / 4 procedures required__ —exception cases of 1 / 2 provided 10 / 15 hours experimental work'.

The most important thing to remember at this stage is to **choose a project that gives plenty of data**. The Investigation Report is marked on a tick-boxes basis and if you don't have enough numbers to work with, it's difficult to be awarded marks in some of the categories. I'm not saying you're doomed if the Report ends up looking like an essay, but an Investigation without numbers from an obviously competent student, is asking a lot from the average SQA marker. The other extreme is to saturate your Report with too much data; you don't want a tedious succession of tables and graphs with hardly any text.

Will your choice require **special equipment**? Your school might have one set of these, or it might be able to borrow it from a neighbouring school. What if it breaks down? Can it resource the purchase of a replacement? In the unlikely event that your school can find the money, can you wait weeks for the delivery? A project using commonly available equipment has its advantages.

Has your chosen project **got enough meat** in it? As the SQA 'suggest' above, it's a good idea for your project to consist of several related experiments, rather than just a single experiment. It makes it easier to fill out the write-up, and avoid the possibility of a project which is a bit 'content-light'. Choose a project with various avenues which can be explored. Not just variations on a theme, but ones requiring different techniques. Google a key word, check out what other people have been doing, then devise your strategy.

Choosing a Project from a List

This is how most students select a project. Schools usually keep examples of Reports from previous years, and if you pick out one that merited a good mark, you will avoid any insurmountable difficulties. The Physics Department may still have some of the equipment that was used; perhaps a huge coil of shellaced copper wire wound on a wooden former, or a transparent plastic tank with a tap fitted to the side. Your teacher has rules governing the use of these old Reports and has to restrict your access to them. Have a good look through it at the initial stages and note things down in your Daybook, but don't expect regular access to it throughout the year. You shouldn't be allowed to take a previous years Report away with you, or regurgitate large sections of it as your own. Yes, get some ideas from it, but add in your own and make it **your** Report, not someone else's.

The next section describes 20 projects from among the SQA's 'most popular'. Read them as 'ways to get you started', rather than 'this is how it must be done'. Add to them!

1. **Pendulums**
2. **Measuring Refractive Index using a Spectrometer**
3. **LCR circuits**
4. **Factors affecting Capacitance and Inductance**
5. **Measuring magnetic Induction using a Hall Probe**
6. **Stretched Strings / Standing Waves**
7. **Charge to Mass ratio of the Electron**
8. **Speed of Sound**
9. **Determination of Planck's Constant**
10. **Interference of Light**
11. **Young's Modulus**
12. **Surface Tension**
13. **Viscosity**
14. **The Wind Tunnel**
15. **Focal Length of Lenses**
16. **The Helium-Neon Laser**
17. **Musical Sounds**
18. **The Bike Wheel**
19. **The Filament Light Bulb**
20. **Lord Kelvin's Bridge**

1   Pendulums

A project measuring the acceleration due to gravity by dropping things has its attractions. My advice is not to do it. A much better alternative is to broaden it out to pendulums. There's plenty of variation, plenty of data, most of it is unfamiliar to the student, and you can still measure 'g'. Only measuring 'g', on the other hand, can tether you to these falling objects and smear you with Standard Grade physics (which, you should know, is a hanging offence).

What most pupils think of as a pendulum, is a weight on the end of a string; wildly flailing uncontrollably back and forth. Boring physics teachers see it differently. It's a small mass on the end of a massless cord, making small oscillations. This is the simplest type of pendulum (hence the name 'simple pendulum'), where all the mass is concentrated at one point and the connection to the pivot point is massless. But there are other types. A uniform metal bar pivoted at one end (or anywhere except the centre of mass), can oscillate with 'pendulum-like' motion. In fact, any shape in a gravity field with the pivot point and the centre of mass in different places, can be a pendulum. These are called 'physical pendulums'; some of which are very cleverly designed.

Start your project with the simple pendulum and do the variation of period with length as your introductory experiment. Be careful to keep the starting angle small and constant, say 15°, and devise a method for doing so. Plot graphs of period (y axis) against length, and period squared against length (expect a straight line here). Make sure you do a thorough uncertainty analysis (see the chapter on uncertainties for handing the uncertainty in the period squared). Hand timing is sufficient for the above and it gives you practice with random uncertainty. Evaluate 'g' from the slope. Investigating the variation of period with mass isn't a good use of your time unless the mass of the cord be comes significant (in which case, it's a physical pendulum). You could investigate the energy loss using a light gate at the bottom of the swing. With each swing, the speed at the bottom decreases (remember potential energy to kinetic energy with a bit of heat energy). Use a cylindrical bob so you don't have

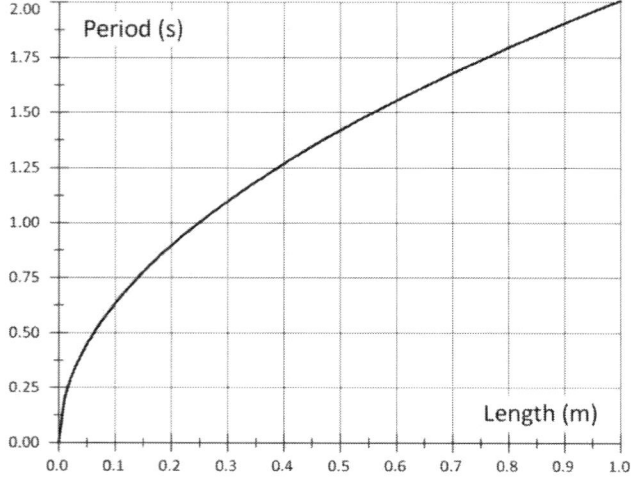
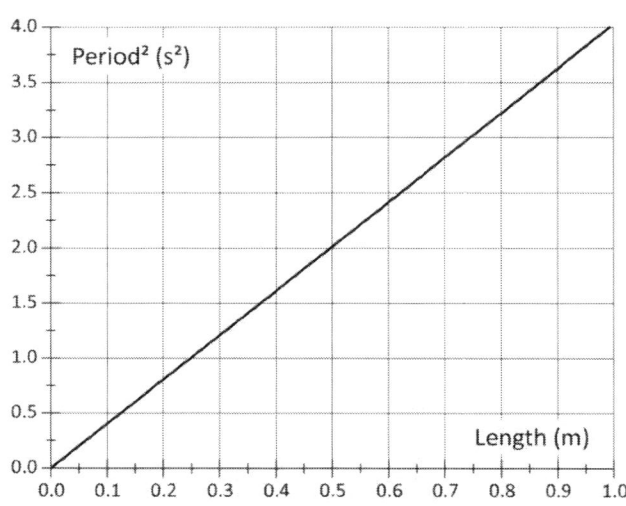

to position the light beam accurately (consider using a laser and stand-alone photodiode rather than the usual orange thing, which you might bash). Record the times for each pass of the bob, measure the diameter and mass of the bob, and calculate the kinetic energy. You could calculate the power loss due to friction from all sources and get an indication of the best type of fixing at the pivot point.

The starting angle isn't supposed to change the period of the oscillation, but it does (slightly). Devise a method for measuring the starting angle accurately. Choose a fixed length of pendulum (say 50cm) and measure the period at 5° intervals of the starting angle. Go up to at least 60°. The expression for the period as a function of starting angle (shown on the graph for 50cm) is:

$$T = 2\pi \left(\frac{l}{g}\right)^{\frac{1}{2}} \left(1 + \frac{1}{4}\sin^2\left(\frac{\theta_o}{2}\right) + \frac{9}{64}\sin^4\left(\frac{\theta_o}{2}\right) + \ldots\ldots\right)$$

Still on the simple pendulum, you could investigate what happens when you use long lengths, like 4m to 5m or more. This would need a convenient (and safe) location. The small angle approximation should be accurate since you're only using 5° angle or less. If you find any variation from the predicted period, check if a heavier mass changes things (compare the mass of string with the mass of the bob)!

If time, try the conical pendulum, but you really want to move on to the physical pendulum; the moment of inertia comes in and the expression for the period

becomes: $T = 2\pi \left(\frac{I}{mgx}\right)^{\frac{1}{2}}$. The length 'x' on the bottom line is the

distance from the pivot point to the centre of mass. The photograph shows an aluminium bar drilled at regular intervals. The pivot point is at the clamp; devise a means of making this as friction-free as possible. An alternative is to use a metre stick, drilled every 5cm, with the holes just big enough to take a smooth nail. The starting angle should be small (about 5°) otherwise the rod starts to flap about. Measure the period for each of the holes on one half of the rod. The complication is that the moment of inertia is different for each hole. You calculate it using the parallel axis theorem: $I = I_{C of M} + mx^2$ with $I_{C of M} = \frac{1}{12}mL^2$ for a

uniform rod. This gives $T^2 = \frac{4\pi^2}{g}\left(x + \frac{L^2}{12x}\right)$. Plot the period squared on the y-axis against

the function $\left(x + \frac{L^2}{12x}\right)$ for a slope of $\frac{4\pi^2}{g}$. Try different shapes. Try Kater's pendulum.

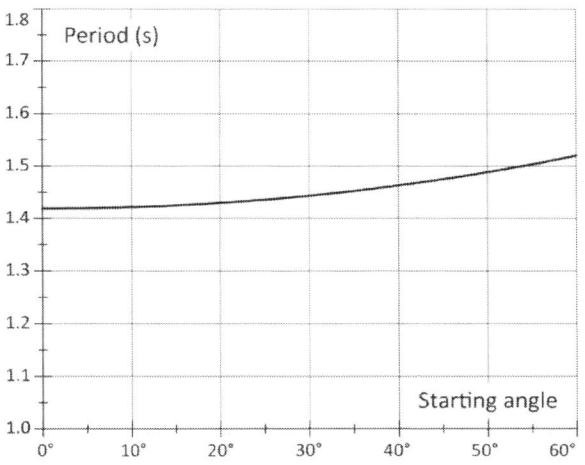

2 Measurement of Refractive Index using a Spectrometer

The temptation is to measure the refractive index of the transparent material using Snell's Law; basically a repetition of Higher work. Even if it's different materials and different shapes, it still isn't enough for Advanced Higher. If your school has a spectrometer, this is definitely the way to go. This is what a basic one looks like:

You might be lucky and have a better one; but they are expensive. It's usually kept in a wooden box and you must take great care of it. Don't lift it either by the collimator (that's the tube on the back left) or by the telescope (the tube in the front right). Get your hand under the base. The small table in the centre can mount a diffraction grating or a triangular glass prism. Your main experiment will use the prism; try to get one which is unchipped (there might be a nice one with the spectrometer). The inset on the lower left is a close-up of the slit on the far end of the collimator tube. If you're lucky, the slit will remain parallel sided as you narrow it. The inset on the upper right is the vernier scale for reading the angle accurately (see page 98 for a discussion on how to read vernier scales). This one isn't decimal; it uses degrees and minutes (hence the 60 on the scale). The reading is 102° 45'.

You should spend a fair bit of time researching and practicing with the spectrometer before using it in earnest. A good description is to be found in R C Stanley, Light and Sound for Engineers, Thomas Nelson and Sons, 1968. For some history, see; http://physics.kenyon.edu/EarlyApparatus/Optics/Spectrometers/Spectrometers.html

Start off your project with the standard Snell's Law experiments using a rectangular block. From an angle of incidence of 5°, go up in 5° stages to 85°. Construct a table like:

Angle of Incidence θ_i	Angle of Refraction θ_r	sin θ_i	sin θ_r
35°±½°	-	0.5736±0.0071	-

Calculate and show the uncertainty for each reading. The uncertainty for the sine function comes from:

$$\sin\theta = \sin 35° = 0.5736 \quad \Delta(\sin\theta) = \cos\theta\,\Delta\theta = 0.819 \times 0.0087 = 0.0071 \quad \Rightarrow \quad \sin\theta = 0.5736 \pm 0.0071$$

Remember to put the $\Delta\theta$ in radians (in the example it's for ½°). No serious rounding at this stage. Now plot a graph of $\sin\theta_i$ (independent variable on x-axis) against $\sin\theta_r$ (dependent variable on y-axis). The slope is $1/n$.

In preparation for the spectrometer experiment, read up on the 'method of minimum deviation' using an equilateral glass prism. With a decent line spectrum for the source, you can measure the refractive index for different wavelengths. You need one that's bright enough (the usual technique is to use a cylindrical convex lens, the type you get in the front of ray boxes, to focus the feeble light along the slit). A good blackout helps; as does a cloth to keep unwanted light away from the prism.

The expression for the refractive index is:

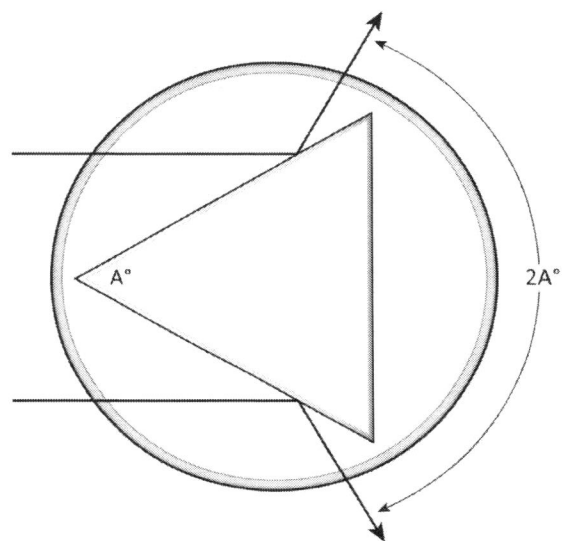

$$n = \frac{\sin\left[\frac{1}{2}(A+D)\right]}{\sin\frac{1}{2}A}$$

You need to measure the prism angle 'A' (don't just assume it's exactly 60°) with its uncertainty (diagram on right). Then measure the angle of minimum deviation 'D' for each spectral line (with its uncertainty). The uncertainty in the refractive index 'n' is calculated using the usual RSS type expression (simplified for A = 60°):

$$(\Delta n)^2 = \left[4\sin^2\frac{1}{2}D\right](\Delta A)^2 + \left[\cos^2\frac{1}{2}(60+D)\right](\Delta D)^2$$

(See the chapter on uncertainties, page 121, showing how to derive such an expression using differentiation).

Next thing is the wavelength of each spectral line. Either look this up in a book, or best of all, use a diffraction grating to measure each one (this would be the basis of a very nice project). Finally, plot a graph of your results with the wavelength on the x-axis. This is called a dispersion curve and the graph on the right shows the result for a typical glass.

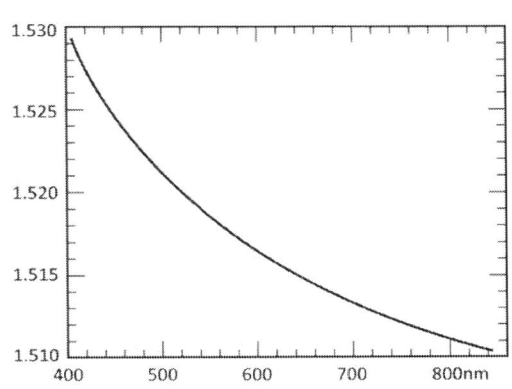

11

3 LCR circuits

This investigation gains a lot if you have access to an oscilloscope which can display two traces on the screen. You'll be using an ac supply and you get phase differences between the voltages across the components. If you don't have this type of oscilloscope (and every school usually has at least one of them), you'll have to make do with a fairly routine investigation on currents in resonance circuits etc. The mains devices you're using (oscilloscope and signal generator) are both earthed so you have to be careful with the connections. Make sure you have a common earth point in the circuit; the black terminal on the signal generator connected to the black terminal on the oscilloscope (usually a bnc to 4mm twin connector like the picture above).

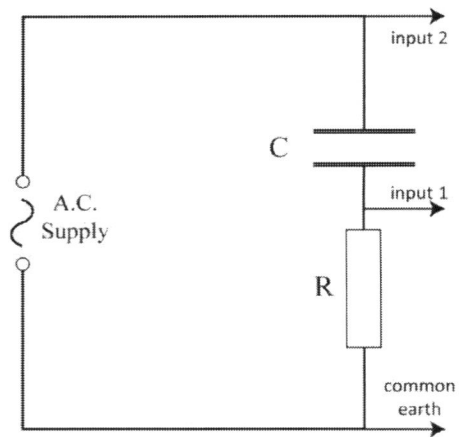

Start with a CR series circuit and ac supply (values to get you started: 2v peak, 8µF, 20Ω). With a twin trace oscilloscope, connect the leads as shown (two leads to the common earth to keep it tidy). Read up about phasor diagrams in ac circuits to understand the analysis. The voltages across the capacitor V_C, across the resistor V_R, and the supply V_o, are related by Pythagoras theorem:

$$V_o^2 = V_C^2 + V_R^2 \quad \Rightarrow \quad Z = \sqrt{\frac{1}{\omega^2 C^2} + R^2} \quad \omega = 2\pi f$$

As connected in the diagram, your oscilloscope will show the voltages V_o and V_R. It also shows the phase difference between these voltages. This depends on the frequency. With the values chosen, you get 45° phase difference at about 1kHz. Set the y-gain to the same value on both inputs, and centre the traces. Measure the peak voltage of both signals and the phase difference between them (V_o will be bigger and lagging V_R). Graph the phase difference against frequency. The equation for the impedance Z above, could be compared graphically with the experimentally measured value $Z = V_o / I$, (use a multimeter in the circuit to measure the current - expect about 50mA rms at 1kHz).

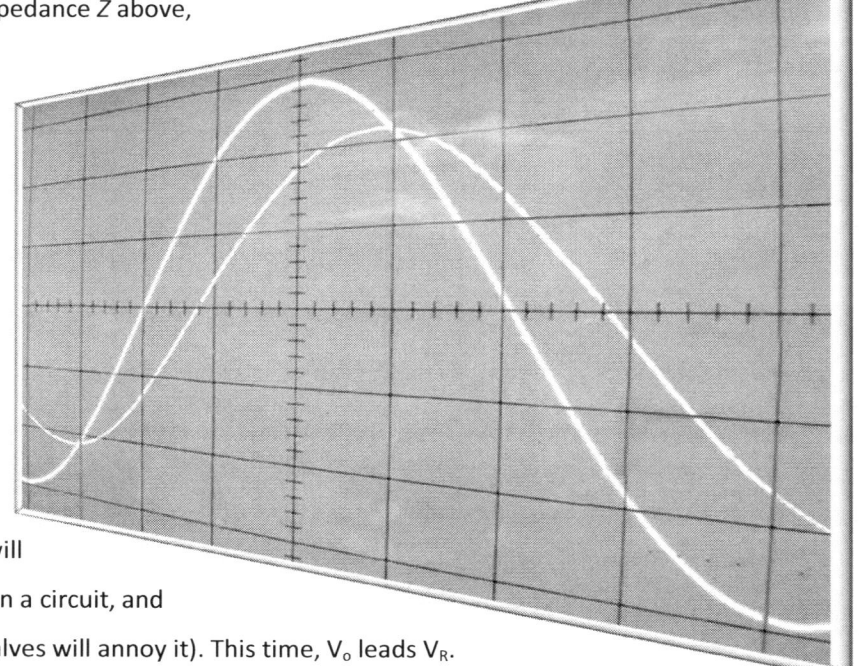

Repeat the above for an LR circuit. I'll leave you to choose values, but just say that inductors are quite 'temperamental' (adding a c-core will dramatically choke off any current in a circuit, and even fiddling about with the two halves will annoy it). This time, V_o leads V_R.

12

Plotting a graph of Z^2 against f^2 (I know it sounds odd) gives a slope of $4\pi^2 L^2$ (to evaluate the inductance) and intercept R^2 (the circuit resistance, which isn't just the resistor). There's plenty of material here for a good 'introduction' and 'procedure' in the write-up.

If you remember about internal resistance from Higher physics, you'll know that the resistance of the supply can affect the performance of the circuit. It's the same with ac supplies. The supply not only comes with a resistance; it's also got a capacitance and inductance. Sooner or later, there effects will show up in your graphs as deviations from the ideal. The details depend on the particular supply you're using, but generally speaking, unexpected inductances become more important at higher frequencies, and unwanted capacitances more important at lower frequencies.

The biggest part of the investigation is with series and parallel LCR circuits. These produce resonance behaviour; as shown below for the series circuit. Though the graph plots the current on the y-axis, it's better

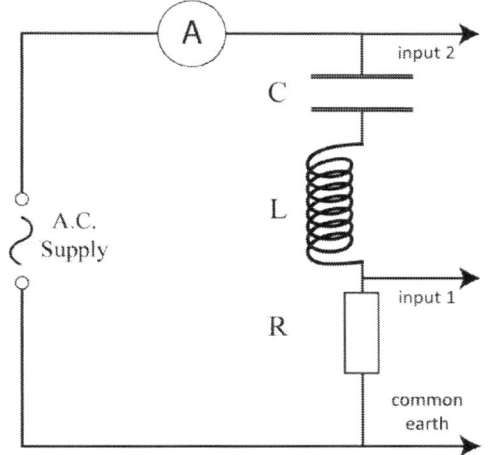
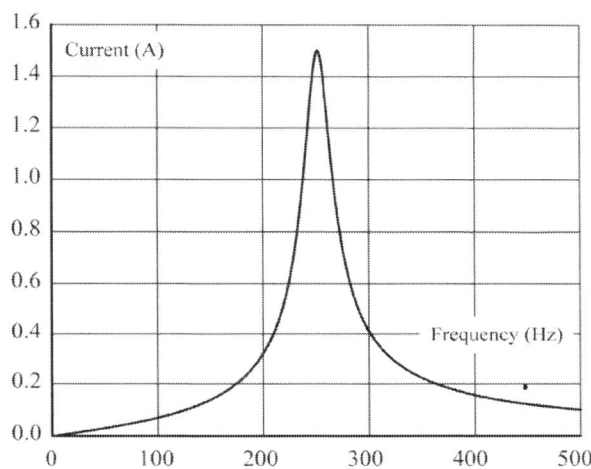

plotting the circuit impedance $Z = V_0/I$, since this is a property of the circuit rather than the supply voltage (and in the process, turn the graph upside down since impedance and current are inversely related). As before with the oscilloscope connections, input 2 shows the supply voltage and input 1 shows the pd across the resistor. The components were, 0.0125H, 32μF, 2Ω, 3V rms. You should investigate the phase difference, especially around the resonance frequency. Repeat for a parallel resonance circuit. The sharpness of the peak depends upon the circuit values; investigate how to get a sharper peak, and relate this to the tuner circuit in a simple radio (remember 3rd year telecomm!).

Further work involves filter circuits. There are several types; the one shown is a band pass circuit. Do some research under 'filter circuits' and see: (click 'next page' for examples)

http://www.allaboutcircuits.com/vol_2/chpt_8/1.html

Filter circuits are useful in HiFi applications: 'cleaning' a power supply by removing unwanted mains harmonics, and in loudspeaker crossover circuits.

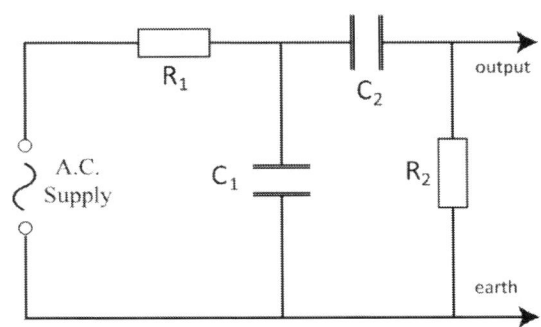

4 Factors affecting Capacitance and Inductance

These are often treated as separate projects; combining them avoids the possibility of a project which is too 'light'. And, on seeing a title like this, the marker would expect experiments below the level required at Advanced Higher. So any way of opening out the subject matter would add greatly to your chance of success. A new title for a start; omitting the word 'factors'. In fact, there are only two factors affecting capacitance (and inductance); the geometry and the materials. Including the 'f-word' in the title limits the scope of the project.

Many schools will have demonstration parallel plate capacitors (about 15cm square or circular, mounted on a base, with 4mm sockets). They are used with an air gap or suitable insulator to show the effect of capacitance on area of overlap, separation, or material between the plates. Charged up, and used with an electroscope, this would be a suitable introductory experiment to show how these factors vary qualitatively (unless you calibrate the electroscope). If you have a nice multimeter for measuring the small capacitance (it'll be in the sub-nanofarad range; multimeters with a resolution of 1pF are available at reasonable price), you don't even have to charge the capacitor yourself.

Another favourite, is to measure the permittivity of free space ε_o using a vibrating switch which charges a pair of plates at a frequency of 50Hz. Redefining the speed of light in 1983 (you no longer *measure* the speed of light), rendered this experiment obsolete, since ε_o will have an exact value ($c^2 = \dfrac{1}{\varepsilon_o \mu_o}$). Look out for these old plates, though, in the physics storeroom; transparent, square, plastic plates about 30cm on a side, aluminium foil carefully flattened on the insides, plus small insulating spacers.

Be careful using capacitors of the electrolytic type. They are soaked in a paper impregnated with a liquid which is not a dielectric, but acts as a dielectric when a voltage is applied. The effect depends upon the polarity of the applied voltage. You must attach the battery to the correct terminals on the capacitor. If you connect them the wrong way to a fairly modest potential difference, the capacitor will produce a gas and may explode or burst because of the pressure inside. I remember deliberately wrecking a capacitor this way; requiring about 20 to 30volts with the particular type used. If the worst happens, don't breathe in the fumes, get everyone out of the room, get the windows open, and get help. They are unsuitable for anything but small alternating voltages. See youTube ('*exploding capacitor*') for a demo.

From your work last year at Higher grade, you may remember discharging/charging a capacitor; and how the time taken depended upon the resistance and the capacitance. There was a 'time constant', given by CR, so if you measure the time constant and the resistance, you can calculate the capacitance.

You have two possibilities. The first is to use a dc supply and choose values so that the time constant works out to be a long time (like more than 10s). This would give you enough time to record the voltage readings by hand in real time (but it will not be accurate), or using a computer interface. The second way uses a square wave supply to continually charge/discharge the capacitor, and either an oscilloscope to monitor the repeat pattern (there's an input on the oscilloscope so that you can trigger the trace at the same rate as the ac supply frequency), or a commercial interface to capture a set of voltage values over at least one cycle. See the circuit on page 12, where the oscilloscope or interface should be connected across the resistor. The graph below shows what to expect on your oscilloscope or through your interface software. The supply is a square wave

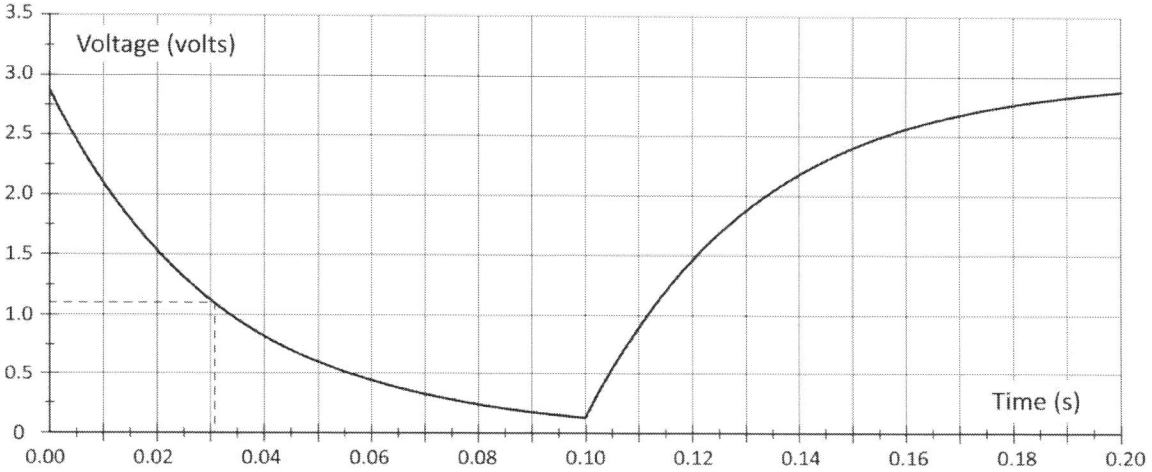

oscillating between 0volts and 3volts (that makes it a dc supply! – an ac square wave stepping between -3volts to +3volts would give a graph with half of it below the time axis). The graph shows a complete cycle of discharge and charge. It takes a time of 0.20s, so the frequency of the signal generator was 5Hz. Here is the derivation for a single charge or discharge:

$$V_S = V_C + V_R \quad \Rightarrow \quad V_S = \frac{Q}{C} + IR \quad \Rightarrow \quad \frac{d}{dt}V_S = \frac{1}{C}\frac{dQ}{dt} + R\frac{dI}{dt} \quad \Rightarrow \quad 0 = \frac{I}{C} + R\frac{dI}{dt}$$

The solution for discharge is: $\quad I = I_o e^{-\frac{t}{CR}} \quad \Rightarrow \quad IR = I_o R e^{-\frac{t}{CR}} \quad \Rightarrow \quad V_R = V_S e^{-\frac{t}{CR}}$

When the time (t) is equal to the time constant (CR), the voltage across the resistor is about 37% (that's $1/e$) of its initial value. It's about 1.1volts for the graph above. This is shown with the dotted line. Knowing the resistance of the resistor was 1kΩ lets you calculate the capacitance. Remember uncertainties.

Repeat with the same capacitor but a different resistance and you should get the same result. Test how high a frequency you can reach with your interface software or oscilloscope display and measure a range of capacitances. Find out about 'stray' capacitances.

Inductors give you plenty of scope for an investigation. The inductance is part geometry and part magnetic properties of the material. With inductors, you can increase the inductance with more turns of wire or by introducing lumps of iron; and there are no nasty chemicals, though you should take care with back emfs (especially with the beast on the right).

There are three ways to determine the inductance. Pass a steady direct current through it and measure the resulting magnetic induction 'B', or switch on a dc supply and measure the slope of the current-time graph at the origin (as in the unit 2 theory section), or pass an alternating current through it and measure its inductive reactance X_L. Coils can be wound without any overlap, or all bundled-up, or a mixture of the two. They can be wound in odd ways, like the toroid mentioned below. This difference in geometry gives them a different inductance (and plenty for you to do). For an air-filled solenoid wound on a non magnetic material, the inductance is given by:

$L = \mu_o n^2 lA = \dfrac{\mu_o N^2 A}{l}$. This is for an infinitely long coil but will be quite accurate if its length is decently longer than its diameter. The small 'n' is the number of turns per metre and the big 'N' is the number of turns. You could simply assume the expression is accurate, measure the number of turns, length and cross sectional area of the coil, and calculate the inductance 'L'. But we're here to do experiments, so check it by passing a direct current through the wire to produce a magnetic induction, $B = \left(\dfrac{\mu_o L}{lA}\right)^{\frac{1}{2}} I$. Use a Hall probe to measure the magnetic induction at the centre of the coil for a given current, and calculate the inductance.

The inductance of a coil can be measured using the ac method. The coil will have a dc resistance 'R' (measure it using an ohmmeter), so use $\dfrac{V_S}{I} = \sqrt{R^2 + \omega^2 L^2}$ to draw a suitable graph (see top of page 13) and calculate the inductance from the gradient. A good scientist would do a rough calculation with the inductance formula to get some idea of the currents and frequencies required in the ac circuit. You could repeat the method for a single Helmholtz-type coil and also for a coil with a c-core (with a big increase in the inductance). There's plenty of scope for students to make their own coils.

If you have time, investigate a type of coil called a toroid. It's like winding the wire around a doughnut, and its inductance is: $L = \dfrac{\mu_o N^2 h}{2\pi} \ln\left(\dfrac{b}{a}\right)$, where 'a' and 'b' are the inner and outer radii of the doughnut and 'h' is the depth of the coil (how wide you have to open your mouth to get your teeth on it). The number of turns is 'N'. If you've more time, look up Maxwell's Bridge.

5   Measuring Magnetic Induction using a Hall Probe

If your physics department doesn't already have a Hall Probe, it's often possible to borrow one from another school. Either way, you have to look after it carefully. The probe sits on the end of a plastic rod with wires running to a box. The box will have a battery in it with an on/off switch. This switch might be spring loaded. If it's not, remember to switch on just long enough to get a reading, otherwise the battery runs down quickly due to the high currents. Getting the Hall Probe working successfully is your first hurdle. You can then use it to measure various magnetic fields (the usual permanent magnets and electromagnets). Revise your theory of the Hall probe from unit 2; in particular that the reading depends upon the orientation of the probe in the magnetic field. Experiment with it using magnets like the one in Standard Grade 'Using Electricity' for demonstrating electric motors (where the field lines go straight between the poles). This will give you the correct orientation of the probe.

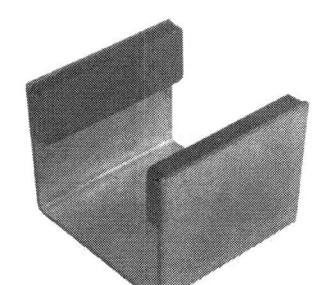

You'll probably have to calibrate the Hall probe; that is, relate the Hall voltage reading on your meter to the magnetic induction 'B'. The simplest way is to use a long (air filled) solenoid like the one below. It's home-made with about 300 turns of enamelled copper wire wound on a plastic pipe (the types used in kitchen or bathroom plumbing like 32mm, 40mm, 50mm, or 110mm for foul drainage are ideal). The magnetic induction along the axis at the centre is given by: $B = \mu_o nI$, where 'n' is the number of turns per metre length of solenoid. Vary the current and produce a calibration curve (hall voltage reading on y-axis, and magnetic induction on x-axis).

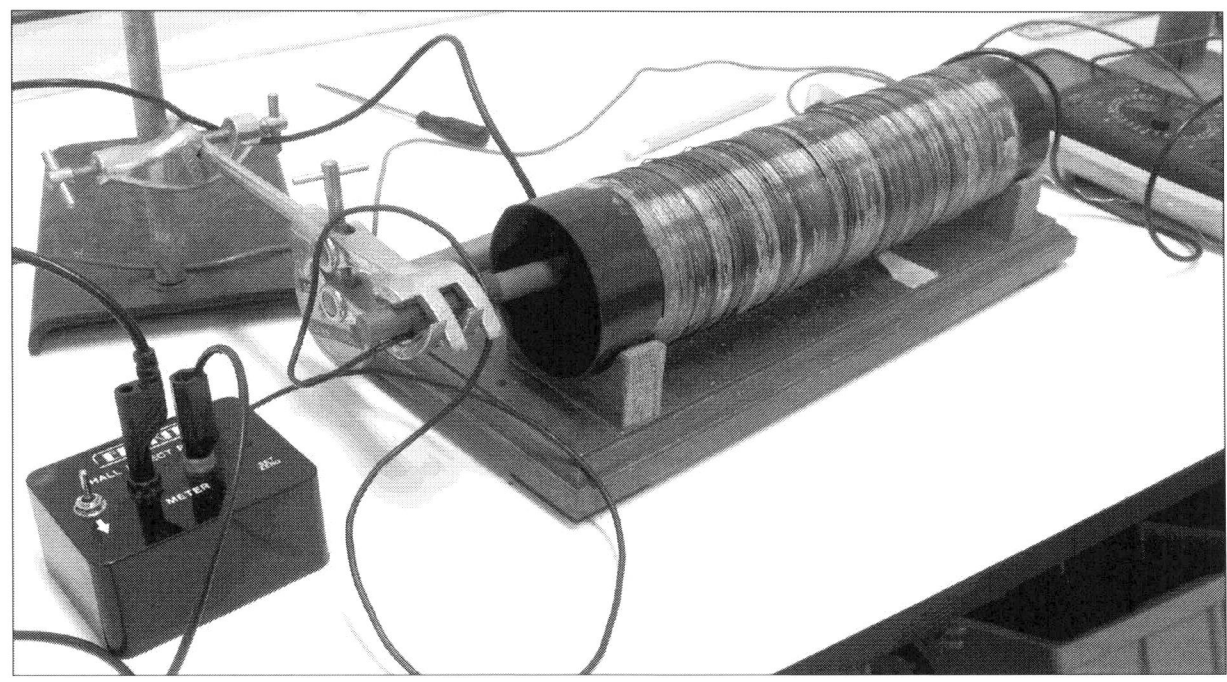

Having calibrated it, you could then measure the magnetic induction along the centre axis, including the region outside the coil. More ambitious still, would be to measure the magnetic induction in the 2D plane. Pay regard to the direction of the field lines (remembering the probe is sensitive to directions). A larger home-made coil will reduce the relative size of the probe. Also try wide diameter, short length coils and check how accurate the previous expression is for the magnetic induction at the centre. This leads on to Helmholtz coils, which should be your main target, providing plenty of 'meat' for your Investigation.

As a first experiment, take a single Helmholtz coil and plot the field along its centre axis. This is like a solenoid scrunched up along its length. The prediction for the magnetic induction at any point along its centre axis is:

$$B = \frac{\mu_o NI}{2} \left(\frac{R^2}{\left(R^2 + x^2\right)^{\frac{3}{2}}} \right)$$

The radius of the coil is 'R' (6.8cm in the ones shown), 'N' is the number of turns (320 in the picture) and 'x' is the distance from the centre along the axis. Measure it using the Hall probe and compare with the prediction from theory using the above equation. For the centre ($x = 0$) compare it with the prediction for the infinite length solenoid $B = \mu_o nI$ (remember the 'n' and 'N' are different things). Don't get the wires too hot!

Now use both coils to make a 'Helmholtz pair'. The idea is to produce a uniform magnetic induction throughout a certain volume of space. This is achieved if the coils are placed a distance apart equal to the radius of one coil. The field along the centre axis is given by: $B = \frac{1}{2}\mu_o NIR^2 \left[\left(x^2 + xR + \frac{5}{4}R^2\right)^{-\frac{3}{2}} + \left(x^2 - xR + \frac{5}{4}R^2\right)^{-\frac{3}{2}} \right]$

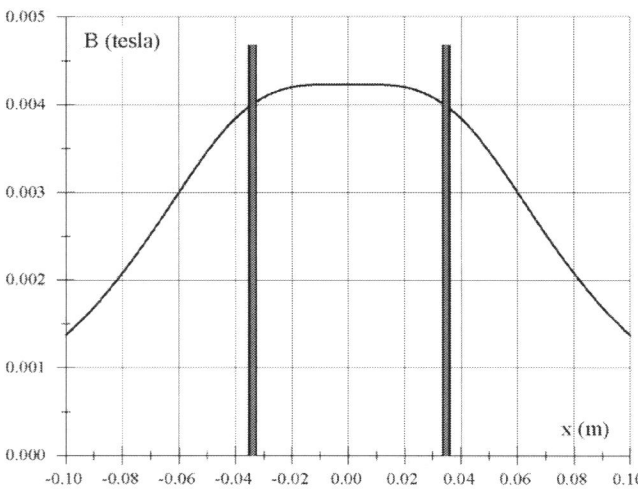

The distance 'x' is measured from the centre and a plot of the magnetic induction against the distance is shown opposite for a current of 1amp. Plot your own graph using the Hall probe and compare with the predicted value from the equation.

Finally, reach a conclusion on whether a uniform field is best achieved inside a long solenoid or between the coils of a Helmholtz pair. Plot both fields on the same graph for comparison.

6   Stretched strings / Standing Waves

Investigating standing waves in a string under tension is an old favourite. There is plenty of data to record, the equipment is easy to use, and there are many variations to explore. An accurate frequency meter is almost essential since the scale on most signal generators isn't usually up to the job. A suitable multimeter with computer interfacing capabilities (like the UNI-Trend model UT70D) can open up possibilities for the more tecky inclined student. The drawing shows the commonest method of producing a standing wave. The waveform on your signal generator should be set at 'sine wave' and the frequency adjusted to the 10's of Hz range. Don't add too much weight to the hanger; it puts a sideways strain on the post of the vibration generator (start with just the hanger). An alternative to the pulley wheel is a knife edge. The pulley wheel is low on friction so the weight of the hanger will almost equal the tension in the string; using a knife edge might produce friction, so the tension in your string will be less than the weight (showing up in any tension graphs as a displaced origin).

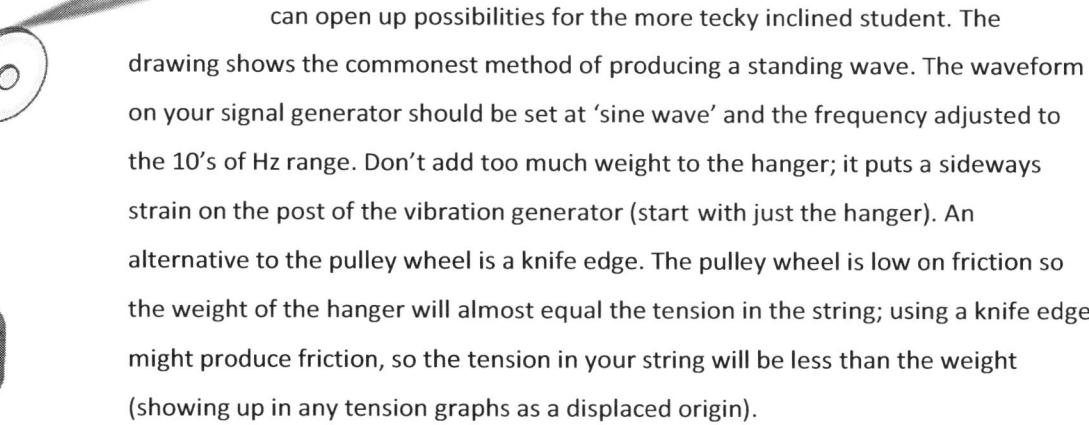

For a given piece of string, the mass per unit length μ (in kg per metre) will be constant. The variables are the frequency of the supply, the tension T and length L of the string, and the number of 'envelopes' n (some people call them 'loops'; whatever you call it, there are two of them to a wavelength). The equation relating them is $T = \dfrac{4\mu L^2 f^2}{n^2}$. You could plot $\{f^2 \; v's \; T\}$ with a predicted gradient of $\dfrac{n^2}{4\mu L^2}$ (graphs for each 'n', and comparison with measured value for μ). The independent variable is the tension, so that's plotted on the x-axis. Using $v = f\lambda$ and measuring the wavelength from the position of the nodes, you can investigate the speed of the waves for one, two, three . . . envelopes.

More difficult, is to measure the displacement of the antinode as a function of frequency. It's quite sensitive to small changes in frequency, so you need a good frequency meter, ideally recording to 0.1Hz. Take readings starting at the lower frequency and increasing it, then repeat the experiment, this time starting from the higher frequency end. Can repeat for 3, 4, 5 . . envelopes; it's a question of how accurate you can measure the displacement of the antinode. You'll need to devise a method of measuring it to at least the nearest millimetre (try a ruler directly behind it, light projection . .).

Other possibilities include: investigating the rotation of the plane of vibration of the string, monitoring the small vibration of the hanger weight due to the string changing the length of its profile (see p35 of the Theory Book for tension complications), or using wire instead of string and inducing standing waves via a current and an eclipse magnet placed at the centre of the wire.

7 Charge to mass ratio of electron

There are two safety aspects to this experiment. One is the use of high voltages and the other is the evacuated tube. Some teachers may feel the experiment is too risky for some students to perform unsupervised; others will deem the risk acceptable (there is a photograph on the St.Andrews University Physics Dept. website showing a first year student without any protection). Obviously, you must take care. Using safety goggles is sensible, and the use of a digital camera to photograph the grid would not only reduce the risk, but also give you a photograph for analysis at your leisure.

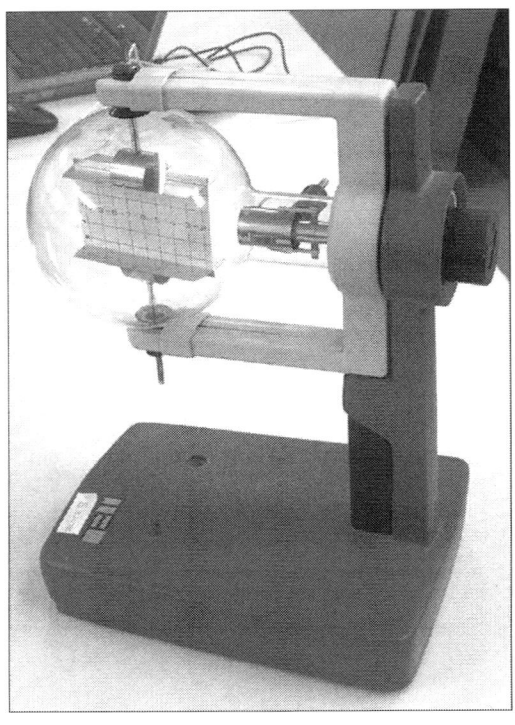

As a single experiment, this would be a bit on the light side for a full investigation, so you should look out for additional material. The charge to mass ratio for the electrons is measured from the parabolic curve of the beam on the screen. The first experiment is to measure the horizontal speed of the electrons using electric and magnetic fields. The magnetic field at the centre of a pair of Helmholtz coils is

$B = \left(\dfrac{4}{5}\right)^{\frac{3}{2}} \dfrac{\mu_o nI}{R}$. Use your unit 2 theory to calculate the horizontal speed of the electrons, $v = \dfrac{E}{B}$.

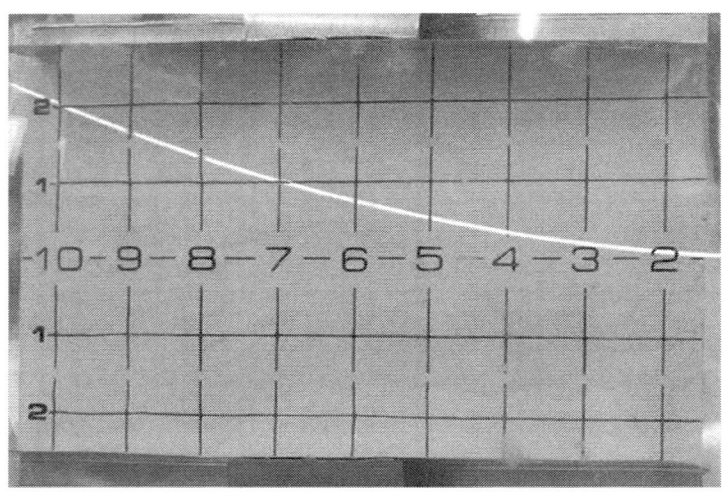

The main experiment passes these electrons between the vertical plates at the grid. The polarity of the plates bends the beam as shown on the left. The equation of the curve is: $y = \left[\dfrac{qV}{2mv^2 d}\right]x^2$ where 'd' is the plate separation above/below the grid and 'V' is the pd. across the plates. You have to extract the q/m ratio from the curve. This is where a photograph comes in handy.

To win over the SQA marker, you have to make as much as you can of the first experiment and do a really good uncertainty analysis throughout. Add in the historical setting; derive all your equations from first principles, and you might end up with a reasonable mark. Have a look at these websites:

http://lpc1.clpccd.cc.ca.us/lpc/physics/pdf/phys2/eoverm.pdf
http://phoenix.phys.clemson.edu/labs/cupol/eoverm/index.html

But it still needs a bit more. How about a Millikan oil drop experiment (and call it 'Properties of the Electron')?

8   Speed of Sound

During the Telecom topic in Standard Grade, you might remember making a sharp sound which travelled between two microphones placed one metre apart. They were connected to a timer which recorded the time taken for the sound to travel from one microphone to the other. Using the distance-speed-time equation gave the speed of sound. You could try this as an introductory experiment; perhaps trying longer distances, more abrupt sound, equalising the amplitude of the inputs to the timer, and so on.

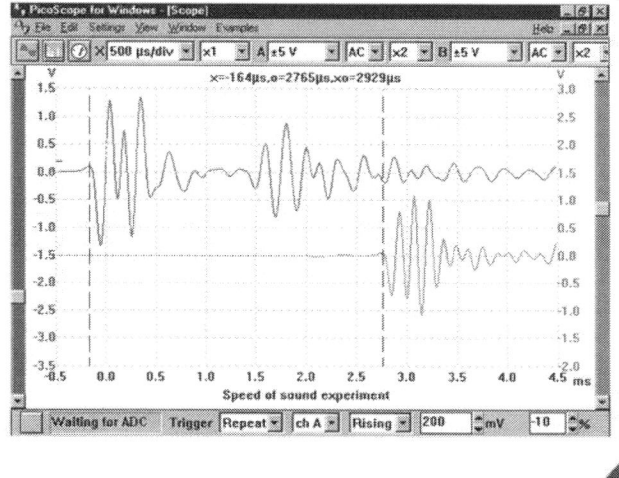

A more up-to-date version uses a PC oscilloscope like the one in the picture. The sound is made as before, but the outputs from the crystal microphones are fed to a box of goodies, and then to the laptop with software to analyse the data. The screenshot shows the signal from the first microphone (top trace) and from the second microphone (bottom trace). The beauty of this is that you can select the points from which to measure, you get a hard copy to play around with, and it has a nice clean tecky feel to it. The downside is the cost (the one shown is about £400, though you can do lots more with it). But, the two microphones technique (even the tecky one), must only be a small part of your investigation. The main part should include one of the methods described below.

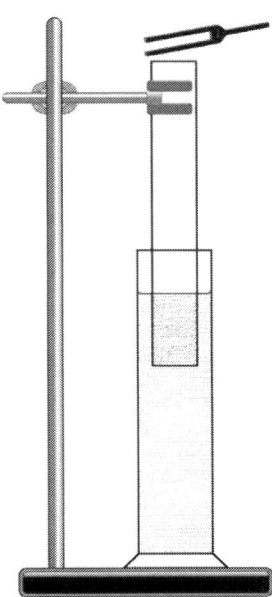

One of the standard techniques is the 'quarter wave tube'. The tube (open at both ends) is placed in water as shown on the left. A pure sound (from a tuning fork or small loudspeaker) goes down the tube and reflects off the top of the water surface. The length of the air column inside the tube is adjusted (by moving the tube or changing the amount of water) until a resonance is heard. At this point, one quarter of a wavelength fits inside the air column. For a 384Hz tuning fork it is about 22cm. 'Resonance' is just another description for a standing wave; in this case it's a node at the water surface and an antinode at the top end. Or almost at the top end. The antinode occurs slightly outside the top end (it's called 'end correction') and you must add this on to the length of the air column for your quarter wavelength. The correction is $0.3d$ and you'll see it depends upon the diameter 'd' of the tube.

An interesting way to measure the speed of sound uses Lissajous figures. Read up on the topic, set up two signal generators and an oscilloscope; familiarise yourself with the basic idea. You normally use sine shaped waveforms, but investigate the squarewave and sawtooth waveforms, and try to make some letters of the alphabet on the oscilloscope screen. It's good fun.

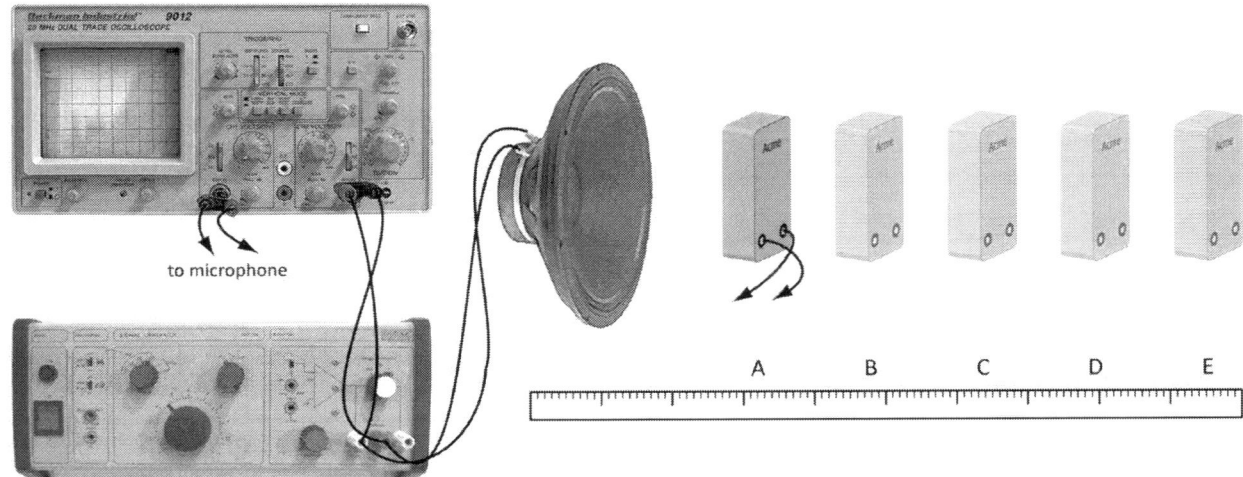

To measure the speed of sound, refer to the diagram above. The basic idea is to emit a pure note on the loudspeaker (with the signal generator also connected to the x-input of the oscilloscope), and detect it with a microphone (connected to the y-input of the oscilloscope). The amplitude of the signal sent from the microphone will be less than that of the signal generator, so you will have to adjust the gains on the oscilloscope to make them about equal (if it's still too small, you'll need an audio amplifier between the microphone and the oscilloscope). Making the amplitudes equal will give circular Lissajous figures rather than ellipses, and straight lines at 45° rather than shallower angles. Locate the microphone at position 'A' where the oscilloscope shows a straight line at 45° (or less if the amplitudes aren't equal). Move the microphone backwards (adjusting the amplitude of the signal to compensate) through positions 'B', 'C', 'D' to 'E'. This is the equivalent of one full cycle (one wavelength). What you should see on the oscilloscope at these positions is shown in the sequence below (don't let the restless trace get you too frustrated).

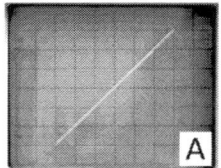 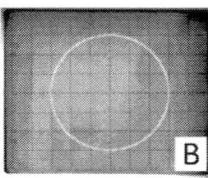 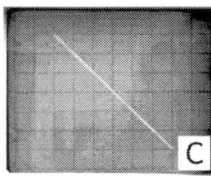 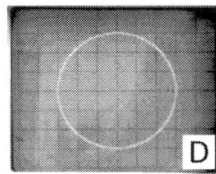 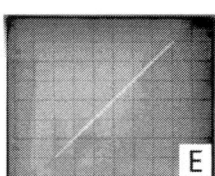

Do a quick calculation using $v = f\lambda$ to get an idea of the scale. For a distance A → E of 0.5m you require a frequency of 660Hz. For analysis, you should measure the intermediate positions (and possibly over two complete cycles or more) and plot a graph of number of waves on the x-axis against the distance on the y-axis. Use the gradient to calculate the wavelength (and remember to do the uncertainty).

Another traditional method for measuring the speed of sound uses Kundt's tube.

9 Determination of Planck's Constant

Planck's constant is the only one in Quantum Theory. It's in the uncertainty relations of energy – time (expressed through the energy of a photon in terms of its frequency) and of momentum - position (expressed through the de Broglie equation). These provide our two methods of measurement.

The first one uses the Planck expression for the energy of a photon $E = hf$, and applies it to a light emitting diode. Obtaining Planck's constant requires the frequency and the energy of the photon. The frequency can be obtained from the datasheet of the LED, or, (and this is what you should do) by measuring the wavelength using a spectrometer. The energy of the photon is equal to the energy needed to produce it through the recombination of the electron-hole pair at the pn junction. This recombination is forced through the application of a potential difference across the junction. As you increase the pd, you'll notice the LED suddenly light up at a certain voltage 'V' (typically a few volts). The work done in recombining the electron-hole pair is the charge moved multiplied by the pd: $\Rightarrow qV$ (where 'q' is the charge of the electron). So, $qV = hf$ and you can work out Planck's constant.

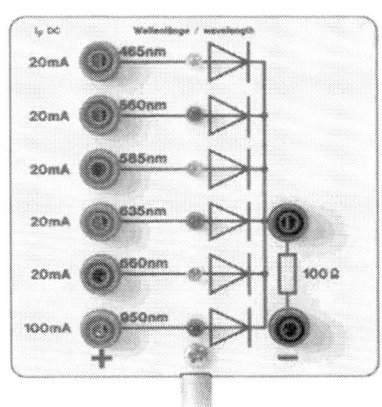

Repeat the experiment with different colours of LED. The photo shows an array of six LEDs, each one emitting a different wavelength of light (from blue to infrared). With components bought in by the school, a student could easily make their own array (the red LEDs cost only about 20p each). Plot a graph of the voltage (on y-axis) against the reciprocal of the wavelength (for a straight line of slope hc/q). Plotting a graph with many LEDs is smarter than using a single LED since it doesn't matter if the line goes through the origin (a single LED calculation assumes that it does).

The actual picture is less clear cut than I've suggested above. Real LEDs aren't monochromatic. The diagram on the right shows the light emission from a gallium arsenide (GaAs) LED. It's about 25nm wide at half height. When you increase the pd across the LED, the first photons emitted are the least energetic ones (the ones with longest wavelength). So if you record the voltage at 'first light', you must take that longer wavelength rather than the one at the peak.

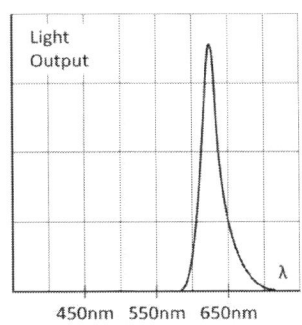

If you look at the LED when it switches on, you'll notice that it doesn't go straight to peak brightness. The bottom graph shows how the current changes (the current is proportional to the light output) with the potential difference for one colour of LED. The minimum voltage (about 1.7v in the graph) teases out the photons of maximum wavelength. That's the one to record.

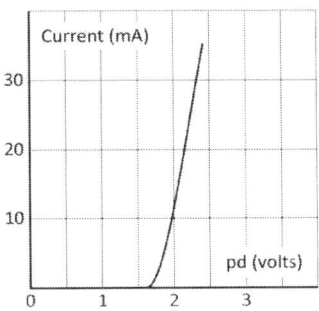

The photoelectric effect also involves photons and electrons. From Higher, you'll remember doing a qualitative analysis showing that photons can behave like particles. Making it quantitative for an Advanced Higher Report gives you Planck's constant. Each photon striking the metal plate gives its energy to an electron to help it escape. Using energy conservation: $hf = W_o + \frac{1}{2}mv^2$. The kinetic energy part is calculated by forcing the electrons to a stop in an electric field (by applying a voltage with the electrons heading towards the 'plus' side – it's called the 'stopping potential' V_o). This gives: $\frac{1}{2}mv^2 = qV_o$. The frequency is the independent variable and the stopping potential becomes the dependent variable; plot the graph for various frequencies of light and measure the gradient h/q. This illustrates the comment on the previous page on the advantage of a graph over a single reading. If you omitted the work function and had assumed that all the photon's energy went into the kinetic energy of the electron, calculating with just one frequency would give you the wrong answer.

The problem with a quantitative photoelectric effect is the availability of equipment. The electrons need a vacuum and that means using a phototube of some kind. The picture above is like the phototube inside the Pasco h/e experiment, though some schools (the type that never throws anything out), will still have them 'loose' (it needs mains voltage across the pins, so look out). But given the equipment, it's a good experiment.

The second main method comes from the wave-particle duality of the electron. You need a Teltron 555 Diffraction tube and not many schools will have one (it costs $900). Read the safety issues concerning these tubes. An evaluation of Planck's constant requires a knowledge of the atomic spacing of the diffracting material (graphite), so it's not as direct as the previous method (which just needs the electron's charge). In fact, the experiment normally assumes Planck's constant and uses it to calculate the atomic spacing! To give the

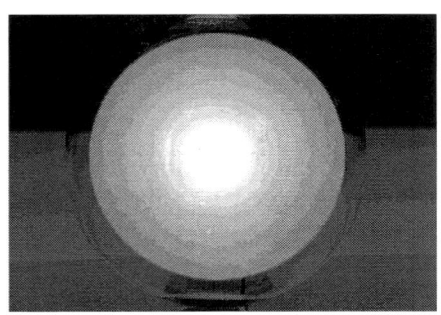

procedure a sound basis, you'd need to derive the atomic spacing from a method independent of Planck's constant. You can do this using Avogadro's number with a density of 2200 kg m^{-3} for the graphite. But this only gives the average spacing of the atoms. Carbon atoms in the form of graphite are arranged in hexagonal sheets, and this gives rise to several regular spacings of the atoms. The ones which produce diffraction rings in the Teltron tube have spacings of d = 0.213nm and 0.123nm, so you get two main rings on the screen. Planck's constant is calculated from: $h = \frac{dD}{L}\sqrt{\frac{1}{2}eVm}$ where 'D' is the diameter of the ring and 'L' is the distance from the target to the screen.

The variables for plotting are 'V' and 'D'. See this website for a good introduction: https://wiki.brown.edu/confluence/display/PhysicsLabs/Experiment+440

10   Interference of Light

This is a good topic for a project; you've covered the basics in your theory units and there's plenty of suitable material for exploring the unfamiliar.

Start with Young's double slit experiment. You can either use a commercially made double slit or one which you make yourself. The photo on the right shows a homemade jig which takes standard sized microscope slides. The slide is painted on one side with a colloidal graphite suspension known as 'aquadag', left to dry, and placed in the position shown. You use a sharp craft knife or fine needle to scribe two parallel lines across the painted surface. There are two ways to do this. Either make them with the knife held at a different angle for each one, or hold the knife absolutely vertical each time and use the screw. You must get the slits as close as possible. It'll take lots of practise, so make up half a dozen slides with ten pairs of slits on each one, then pick the best. You can tell which one is the best by shining a laser on it and looking at the results on a screen.

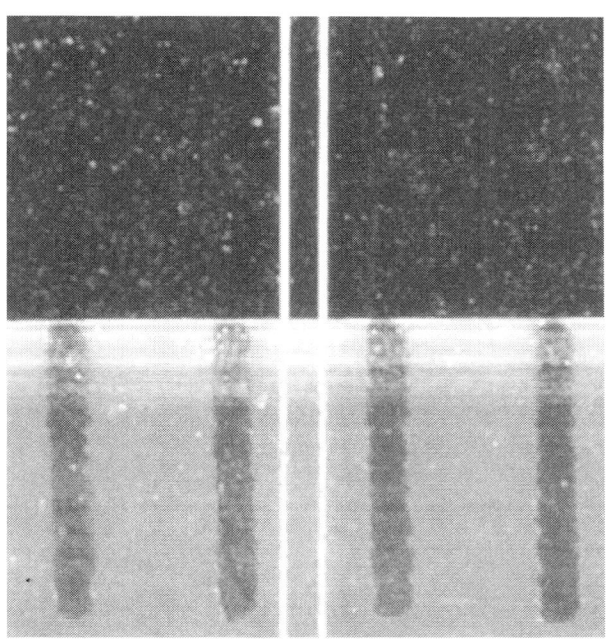

For analysis you need to measure the centre to centre spacing of the slits (the width of each slit is also useful). The traditional method is to use a travelling microscope, but the advent of modern digital cameras offers another method. The photo on the left, of a double slit beside a plastic ruler, was taken with a moderately priced digital camera using the x16 zoom function on a manual setting. Download it into your computer, manipulate the image with some readily available software, and you have a hard copy for measurement at your leisure (and a nice picture for the Report). You can see from the picture that it's also possible to estimate the width of a single slit; something which is almost impossible with the travelling microscope.

You can begin by repeating the standard experiment to measure the wavelength of the laser (it's 632.8nm), but your main experiment should measure the intensity variation of the pattern on the screen. The fringes should look like the ones at the bottom of the page. To measure them requires the ability to precisely position a 1mm^2 photodiode. Refer to the experiment on lasers (page 39) for details of a suitable homemade device. Notice how the fringes fade out on either side. This is due to the diffraction effect of the width of each slit. If you have several decent pairs of slits, perhaps all of the same slit width but different slit separation, you can investigate

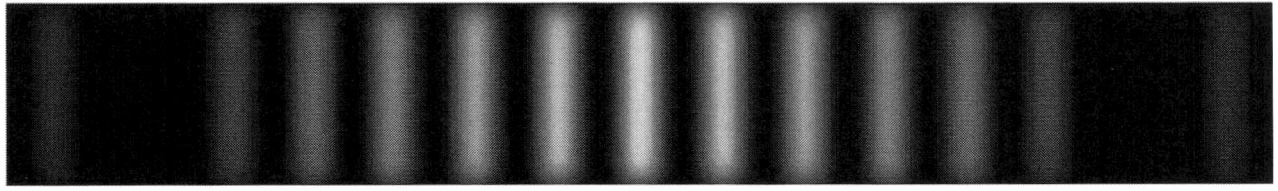

this effect. Width to separation ratios are typically 1:4 (like the one in the photograph), up to 1:10. If your maths is up to it you can compare your result with the theoretical prediction for the intensity:

$$I \propto \frac{\sin^2 A}{A^2} \cos^2 B \quad \text{where} \quad A = \frac{\pi a \sin\theta}{\lambda} \quad \text{and} \quad B = \frac{\pi d \sin\theta}{\lambda}$$

The slit width is 'a', the slit separation is 'd', and 'θ' is the same as in unit 2 theory.

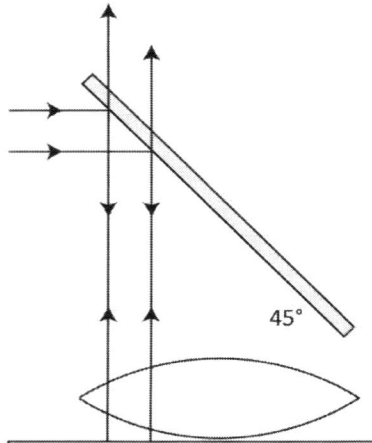

Moving on to interference by division of the amplitude brings us to Newton's Rings (do the air wedge first if you need even more to do). The basic apparatus is a thin convex lens (or plano-convex lens) sitting on top of a flat glass plate, illuminated by sodium light (don't put white paper underneath the plate). A microscope slide is positioned at 45° to keep your face in a different place from the light source. A series of rings is seen with a dark spot in the centre; the diameter of the nth dark ring given by:

$d_n^2 = 4n\lambda R$ where $n =1$ for the first dark ring etc. The 'R' in the formula is the radius of the curve of the convex lens. Plotting a graph of d_n^2 v's n will give a straight line of slope $4\lambda R$. Measuring the radius isn't as easy as you might think. Most physics departments will have a spherometer like the one in the picture on the right (hidden away in a small, dog-eared cardboard box in the back of a drawer). Try it out, and be prepared for some frustration. An alternative is to use light projection. Perhaps this is why Newton's Rings is used to measure 'R'?

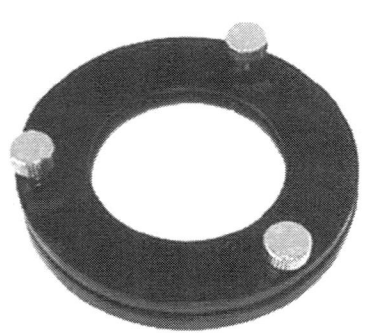

There is another method for measuring Newton's Rings where you observe the light transmitted through the system rather than the light reflected back. This uses a special mount (used in the upright position), with three adjusting screws, and is shown in the picture on the left. Try making one yourself if the physics department doesn't have one. The screws hold the two pieces of glass in position and also allow adjustment of the plano-convex lens to make the rings circular. The image on the screen shows a bright spot at the centre this time (rather than a dark one as with the other method) due to an additional reflection from the underside of the convex surface. The equation is just the same as the previous one; this time measuring the radii out to the bright rings. The advantage of this method is that you don't strain your eyes looking through a microscope, though you must devise a method for calibrating the scale on the screen, and the contrast isn't as good. Try this website for some details:

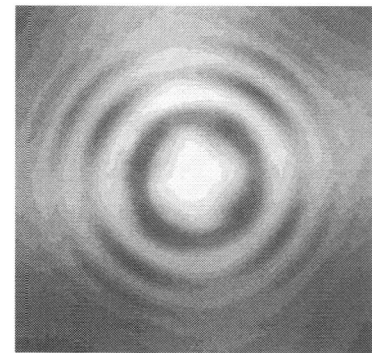

http://www.nikhef.nl/~h73/kn1c/praktikum/phywe/LEP/Experim/2_2_02.pdf

Many schools will have a set of six commercial 35mm size slides of slits ranging from a single slit up to six slits. These provide ample opportunities for plotting intensity patterns and comparing with theory.

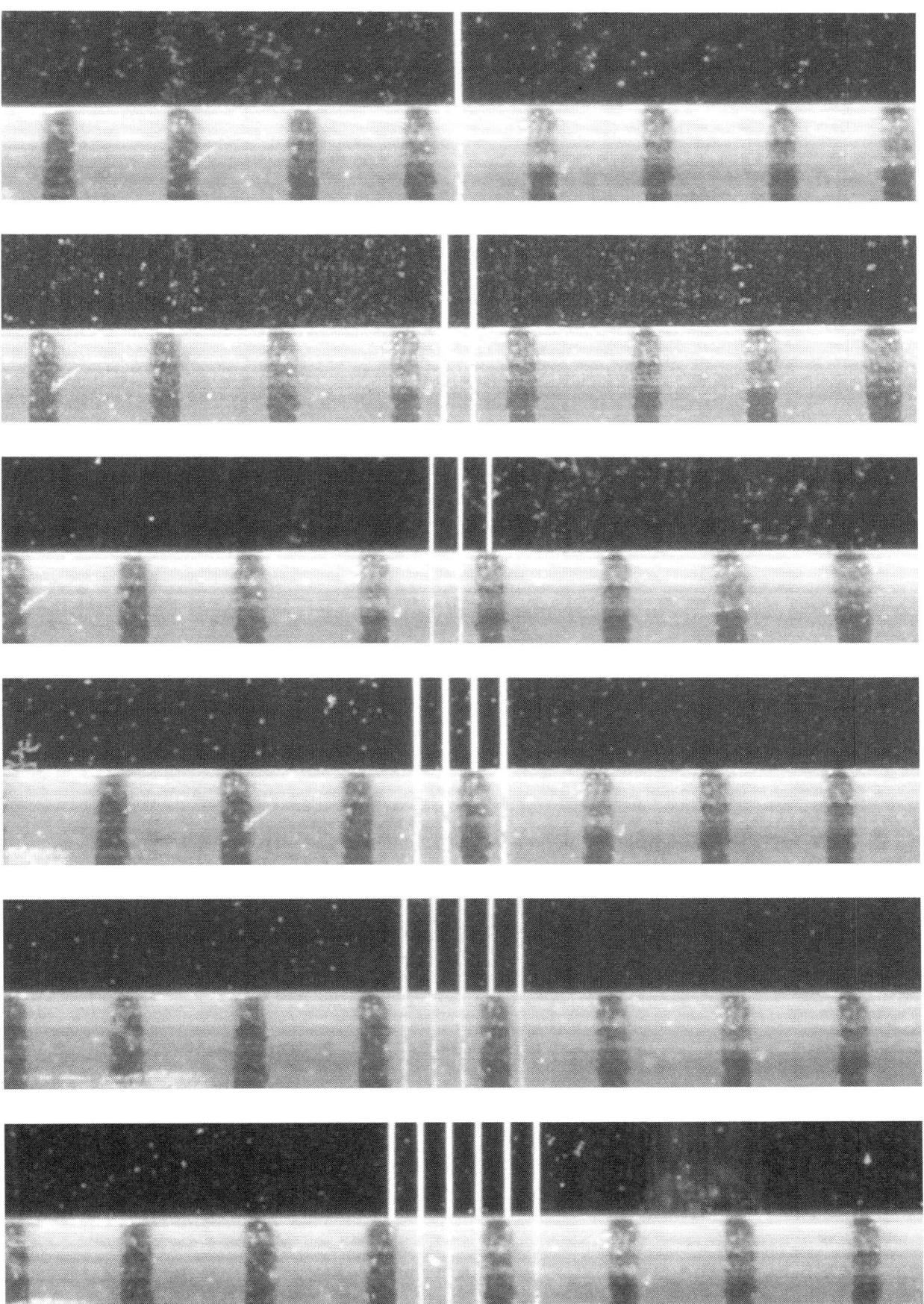

11   Young's Modulus

I always get the feeling that there's too much engineering and not enough physics in this project. In that respect it shares the same fault as some electronics projects ('it's not quite physics'); but it is popular, there's no shortage of data, and you won't get bushwhacked by difficult concepts.

Young's modulus (it's the same Young from double slit interference) is a way of quantifying the 'stiffness' of a material which is independent of the shape. This is why its definition contains force *per unit area* F/A and the *fractional change* in length $\Delta L/L$, otherwise known as the stress and the strain. There are three standard experiments for measuring Young's modulus, each one applying a force to change the shape, and each one requiring measurements of that shape.

The first experiment stretches a wire. It's that simple. You just have to be careful not to injure yourself; things like heavy weights falling on your foot, breaking wires whipping your face etc. Be sensible and take precautions. You have a choice between a vertical wire (attached to a clamp stand at the top and a weight at the bottom), or a horizontal wire like the arrangement shown below.

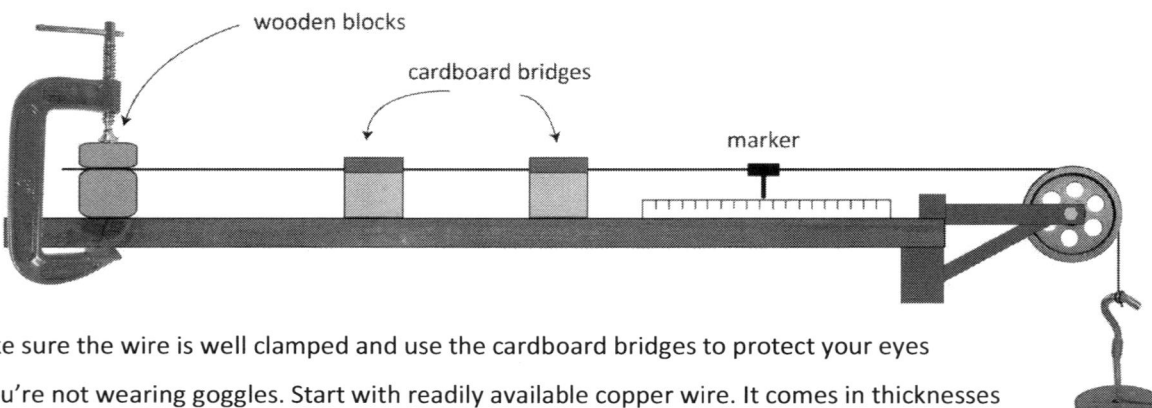

Make sure the wire is well clamped and use the cardboard bridges to protect your eyes if you're not wearing goggles. Start with readily available copper wire. It comes in thicknesses labelled by a 'gauge number', like swg28 or swg36. The bigger the number, the thinner the wire. Google 'standard wire gauge', for the sizes. You record the weight and measure the increase in length. The diagram shows a marker system, and if you place it in the position as shown, the length of wire is the distance from the clamp to the marker. You can easily devise a better system (especially since the increase in length is small). The expression for Young's modulus (the one with the *ΔL* below) shows that the increase in length is proportional to the force. This means; keep within the elastic limit for a constant Young's modulus.

Young's modulus '*E*' is calculated using: $E = \dfrac{F/A}{\Delta L/L} \Rightarrow \Delta L = \left(\dfrac{L}{EA}\right)F$. Plot a graph of the force applied to the wire (on the x-axis) against the increase in length on the y-axis. The gradient will be $\dfrac{L}{EA}$. Remember to use SI units throughout, with conversions from mm^2 into m^2, to get an answer somewhere about 120GPa (pascals is a surprise, but it's to do with energy and volume). Repeat for different thicknesses and different materials.

The second method applies a force to the centre of a regular beam supported at its ends. As before, you don't want to bend the beam out of shape. Just apply enough weight to make it sag, then recover, when you remove the weight. The expression for the downwards deflection 'd' is very like the previous one; only the bit inside the bracket changes. Devise a method for measuring the deflection, and don't make the bar too thick. There are two versions; one for a bar of circular cross section, and the other for a bar of rectangular cross section.

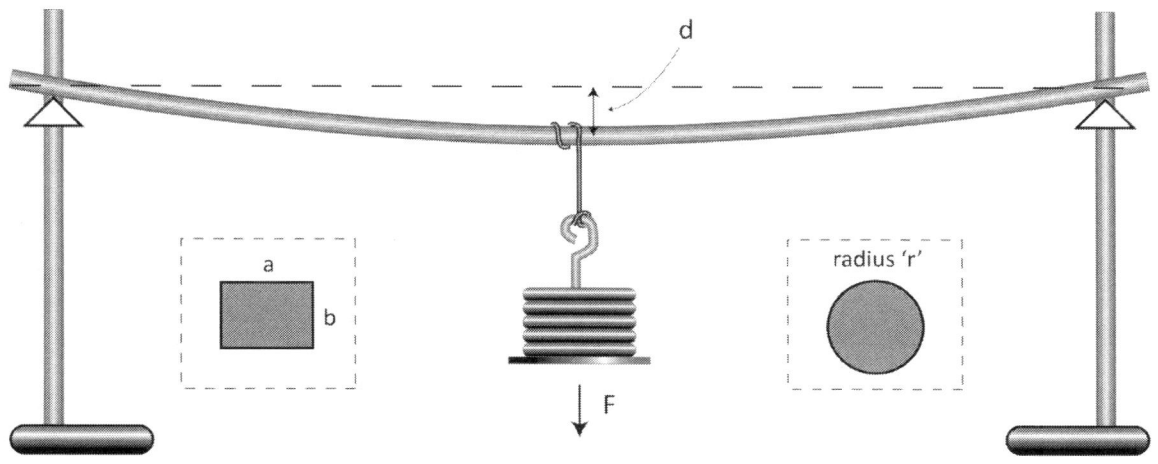

For the rectangular section: $d = \left(\dfrac{L^3}{4ab^3 E}\right) F$ For the circular section: $d = \left(\dfrac{L^3}{12\pi r^4 E}\right) F$ Plot as before.

The third method is called a cantilever. One end of the rod is firmly clamped to the bench ('encastre' as it's called) while the other end is loaded with the weights. The deflection is recorded where the weights are applied. There are two things to look out for. Like the previous experiment, if you're using a bar of circular cross section, make sure it doesn't roll anywhere. Take steps to prevent this. And in this experiment, make sure

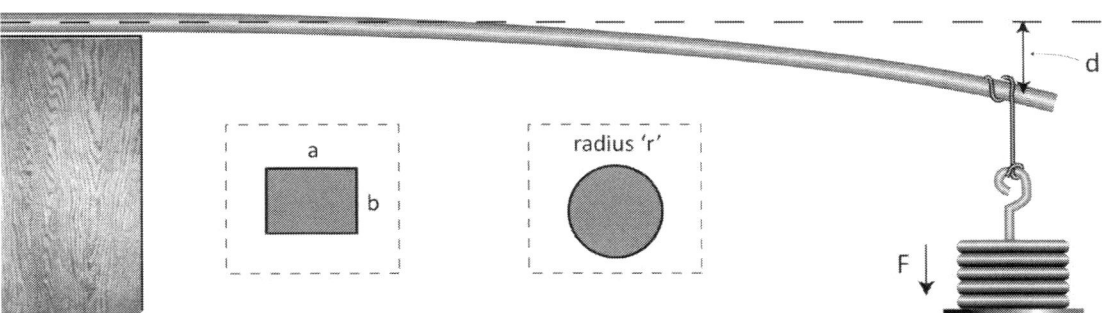

the weight doesn't slip off the end of the bar. Remember to remain within the elastic limit and not to select a bar which is too thick (using the formulae below, and taking Young's modulus for steel as about 200GPa, should give you an idea of what to expect).

For a rectangular section: $d = \left(\dfrac{4L^3}{ab^3 E}\right) F$ For a circular section: $d = \left(\dfrac{4L^3}{3\pi r^4 E}\right) F$ Usual graphs.

The project gives plenty of data, plenty of uncertainties, and a straightforward write-up. It just lacks variety.

12   Surface Tension

Many students develop an initial interest in physics from watching TV programs on black holes or particle physics; or just looking up at the night sky. I've yet to meet anyone who got started with surface tension. It's not in the syllabus. Doesn't even come near the syllabus. At once, it's both messy from its mystery, and inviting for its many avenues (or is it the other way around?). But give it a try. It's a fascinating topic when you get to know it, with many unexpected applications in everyday life from GoreTex jackets to wetting agents.

With the Report in mind, the project can be a bit short of graphs unless you explore temperature variations, or manage to get many widths of capillary tubing, or a sympathetic marker. Start off with reading-up on surface tension (the internet is best for a quick survey). Try to understand what the molecular forces are all about, the difference between adhesion and cohesion, the importance of the material the liquid is resting on, the angle of contact, and why the units are newtons per metre. Here are three methods for measuring surface tension.

Gather together a collection of **capillary tubes** with different internal diameters (either side of one millimetre). For a given bore of tube, the greater the surface tension, the greater the height it rises to. The height 'h' is also

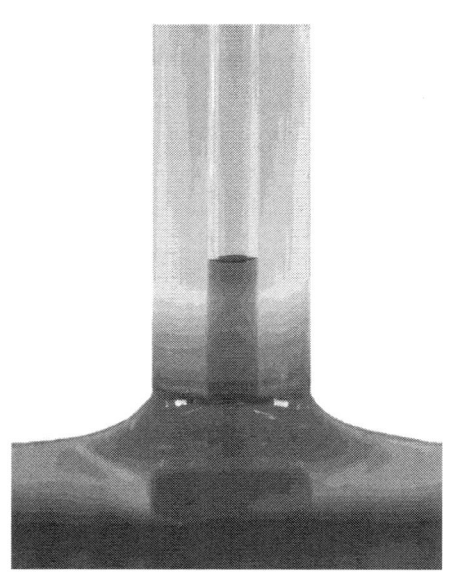

inversely proportional to the diameter 'd' of the liquid column. This provides your chance for a graph. Starting with distilled water, set up the capillary tubes (they need to be clean inside; spend time on this because it's very important) and measure the height of the liquid in each one. Measure the internal diameter of each tube using whatever method you can think of for the wider ones, and a volume method for the narrower ones (work out for yourself, then describe in your report, how this gives a small fractional uncertainty in the diameter). Plot the inverse of the diameter on the x-axis against the height on the y-axis. The equation is:

$$h = \frac{4\gamma \cos\theta}{\rho g d}$$

The equation contains the density ρ of the liquid, the contact angle θ, and the surface tension γ (in units of N m^{-1}). Strictly speaking this should be called the surface tension of the liquid against glass γ_{LG}. Recall the shape of the cosine graph from maths (it's positive up to 90° then negative to 180°), research the correct definition of the angle (where 0° starts from), then explain why mercury has a convex meniscus and where the level would be in a capillary tube. Evaluating the surface tension γ requires you to measure the angle (try a digital photograph).

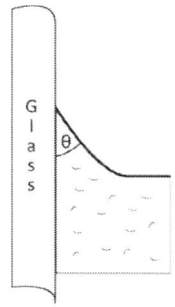

The second method is called the **sessile drop**. A drop of liquid with no surface tension will lie flat on a solid surface (making an angle of 0°). A drop which sits on the surface as a perfect sphere will contact the surface at one point. The tangent at that point is parallel to the surface and makes an angle of 180°. Real

drops take a shape somewhere between these two extremes, like the one in the photograph (they can also slump over time). The height of the drop is given by: $h^2 = \dfrac{2\gamma(1-\cos\theta)}{\rho g}$

You'll need a good photograph of the drop with a millimetre scale in the view. As before, the surface tension should really be written as γ_{LS}, since a water drop on paraffin wax is quite different from one on fused quartz.

The third way is called Du Nouy's method. It's based on a length of wire formed into the shape of a circle. Make it about 5cm in diameter from wire about 1 to 2mm thick (swg 14 to 18), and make sure the circle is perfectly

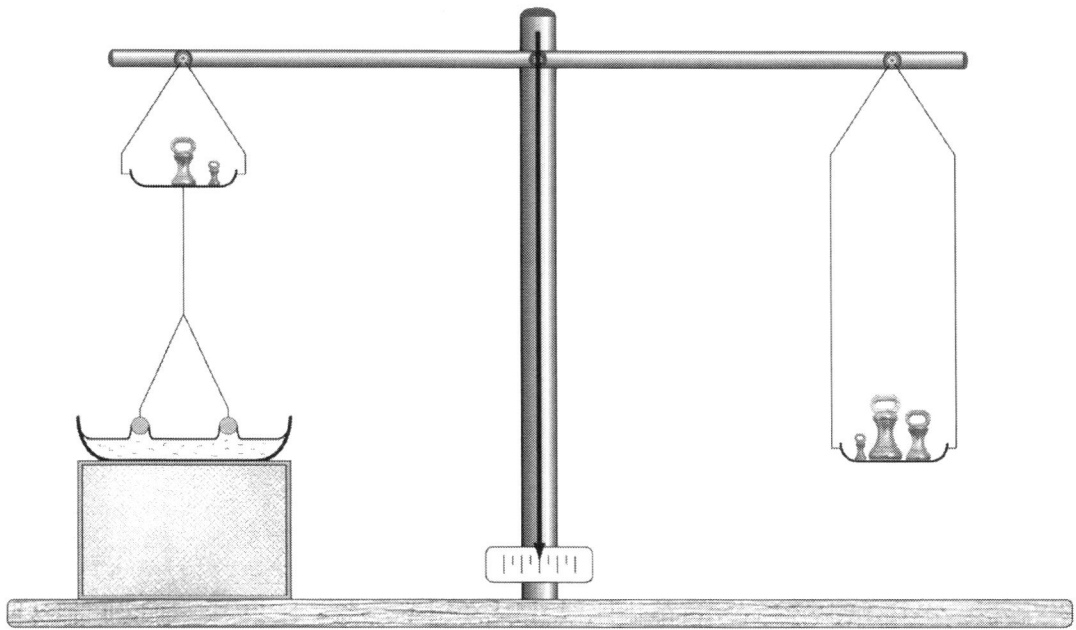

flat. You set up the equipment as in the diagram, but with the circle of wire below the surface. Get the system balanced at this point. Add small masses to the right-hand pan and the wire circle will move upwards. When it reaches the surface, the effect of surface tension will try to keep the wire in contact with the liquid. Record how much extra weight ΔW is required to make the wire just break free of the surface (remember to keep the wire horizontal). The surface tension is calculated using: $\gamma = \dfrac{\Delta W}{2\pi d}$ where 'd' is the diameter of the wire ring (look into why there's a '2' on the bottom line). Success depends upon a properly cleaned ring and no upsets like vibrations. Attach the wire ring using three threads. Consider investigating the variation with temperature.

13 Viscosity

Like surface tension, viscosity isn't in the syllabus. Mentioning it at a party won't make you the centre of attention. In fact, it might have the opposite effect. Depending on the party, you could enthral them with descriptions of lava flow from volcanoes, the blood flowing through their blocked-up arteries, or the hot oil pumping through the engines of their 4x4's. And if you look like a Hollywood film star, you might even get away with it for a while; but sooner or later you'll hear the sound of pens scraping over your name in their address books. Ok, it lacks a certain social cachet. So what. When you're up for that job interview and asked what project you did in sixth year, you can tell them something that's different, and something that's useful. You're also that bit more employable.

The physics of viscosity is probably new to you, so I'll begin with a brief overview of the topic. It's part of a larger area called fluid mechanics, and includes gases as well as liquids (which is why you apply Stokes Law in the Millikan oil drop experiment). The basic idea is to consider a fluid to be made of many layers. Apply a small force to the top layer (along the direction of the layer) and keep the bottom layer at rest (perhaps it's up against something solid). The force of 60N on the top layer acts upon a 2m² area of surface. This is a stress of 30 Nm⁻². The effect it has due to the 'stickiness' of the layers against each

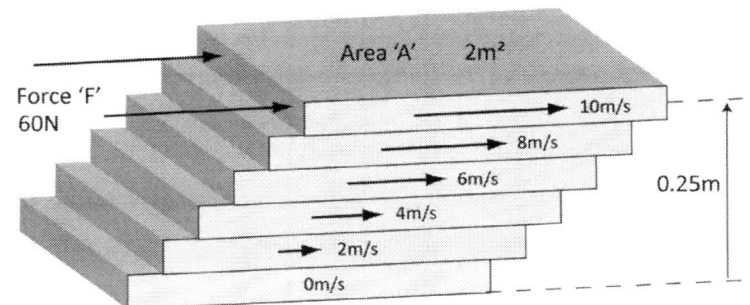

other, is shown with the speed of each layer. The speed changes from rest at the bottom layer to 10 ms⁻¹ at the top layer over a distance of 0.25m. This is a velocity gradient $\frac{dv_y}{dz}$ of 40 ms⁻¹ per metre (y-axis goes to the right, z-axis is up the way). This dragging out of the speeds is proportional to the force applied to the top layer. Or at least it is in some fluids; they are called Newtonian fluids. For these fluids: $\frac{F}{A} \propto \frac{dv_y}{dz} \Rightarrow \frac{F}{A} = \mu \frac{dv_y}{dz}$. The constant of proportionality μ is called the **dynamic viscosity**, and from the equation its units are kg m⁻¹ s⁻¹ (it's the same as pascals times seconds, so the common unit is Pa s). For the diagram $\mu = 0.75$ Pa s. This is slightly less than the value for glycerine (0.934 Pa s at 25°C) but much greater than water (0.001 Pa s at 20°C).

Now for the complications. The above has nothing to do with gravity, but many experiments make use of gravity; such as timing a viscous fluid draining through a narrow bore tube. In these experiments, there is a second way of defining viscosity, and it's called the **kinematic viscosity** v, with units m² s⁻¹. The classic example illustrating the difference, compares a jar of honey with a jar of water. A non-gravity method such as measuring the viscosity by stirring it, would show the honey to have a larger (dynamic) viscosity. A gravity method such as laying the jars on their sides and seeing which drains the fastest, would show the honey to have a larger (kinematic) viscosity. For Newtonian fluids the two are related. Divide the dynamic viscosity by the fluid density to get the kinematic viscosity. With many liquids having a density either side of one thousand (in SI units), the two answers will be a lot different. For water $\mu = 1 \times 10^{-3}$ Pa s and $v = 1 \times 10^{-6}$ m² s⁻¹.

The second complication is the difference between laminar flow and turbulent flow. The last diagram showed the water behaving in a nice regular way. There were no eddy currents. When an object moves faster through a fluid, the flow pattern can change from laminar to turbulent, and the way to calculate when this happens uses Reynolds number 'R'. Work out the combination $R = \dfrac{\rho v r}{\mu}$. This is for a sphere of radius 'r', with a speed of 'v' in a liquid of density 'ρ' and dynamic viscosity 'μ'. If it's about one or above, you loose laminar flow and cannot use equations like Stokes Law. You'll need Stokes law for the experiment where you drop steel spheres in glycerine, and with the radius of the sphere on the top line, you cannot use too big a diameter. For a Reynolds number of one, the maximum diameter of sphere is calculated from $D_{max} = \left(\dfrac{18\mu^2}{g\rho_{fluid}(\rho_{obj} - \rho_{fluid})} \right)^{\frac{1}{3}}$. For a steel sphere in glycerine, it's about 6mm diameter. Bear this in mind when you do the experiment.

Here are three common ways of measuring the viscosity of a liquid. They're all gravity methods, so strictly speaking it's the kinematic viscosity we're measuring. The equations are expressed in terms of the dynamic viscosity 'μ' (assuming the density relation between them).

The first one uses what's called Poiseuille's Law. A viscous liquid runs through a narrow bore pipe and is collected in a tub. How long it takes to do this is related to its viscosity. The expression is: $\Delta P = \dfrac{128 \mu L Q}{\pi d^4}$. There's the difference in pressure between the top of the liquid in the reservoir and the pipe outlet ΔP, the length of the narrow bore pipe (L), the diameter of the narrow bore pipe (d), and the flow rate out of the pipe (Q) in metres cubed per second ($m^3\,s^{-1}$). Be careful with the units.

The pressure difference can be calculated using your expression from Higher Physics $\Delta P = \rho g h$, where the height is from the top of the water down to the narrow bore pipe (will this be accurate if the long pipe isn't horizontal?) Evaluating the viscosity 'μ', is easier if the flow rate of the liquid is constant, and to do this, you must keep the pressure constant. The diagram shows the simplest way to maintain a 'constant head' of pressure. The bent tube is connected to the water tap and the excess runoff goes to the sink. However, this limits you to water as the liquid. You should also try motor oil (there are different SAE grades), but using your own method to maintain constant pressure. Try different bores of pipe with water, graph the flow rate against the fourth power of the diameter and measure the gradient. Best of all, try to measure the variation of viscosity with temperature for motor oils (that's why the pipe is shown with insulation). This is very important quantity in engine wear.

The second method uses a U-Tube viscometer like the one on the right. The basic idea is to time how long it takes the viscous fluid to drain away through a narrow bore pipe, so it's similar in method to the previous technique. The traditional method for starting is to syphon the liquid up into the upper reservoir. (Notice the glass connection bridging the upper arms of the U-Tube. Do you think it's solid or hollow? What is its purpose?) The upper bulbous reservoir has two marks, one just above the reservoir (at A) and one just below (at B). When ready, start the timer as the liquid level drops to the 'A' mark and stop the timer when it reaches the 'B' mark. That's all you have to do. There is no equation to calculate the viscosity. You simply look up the calibration table or graph which came with the viscometer when it was bought. That doesn't amount to a lot of physics for a Report. However, these tables will be based on a dynamic viscosity of water at 20°C of 0.0010016 Pa s, so if you can't find your calibration data, you could perform an experiment with water at 20°C and at least get one point fixed. The viscometer can be placed in a suitable water bath for investigating temperature variations, which would at least give you a graph of temperature (x-axis) against time to drain (y-axis). Keep it clean.

The third method involves dropping spheres into the viscous liquid. It's different from the last two. This time, the liquid doesn't move; the solid moves through it. Use steel spheres with a range of sizes (but remember the Reynolds number story). The liquid has to be quite gooey; glycerine is about right, but you'll need quite a lot of it. Another possibility is wallpaper paste, where you could investigate its viscosity as a function of the concentration of the powder. Either way, it's messy. You've got to spend time cleaning equipment, and things like how to retrieve the spheres from the bottom of the cylinder. The viscosity comes from:

$$v_{term} = \frac{2r^2 g}{9\mu}\left(\rho_{obj} - \rho_{fluid}\right)$$

It's got the radius of the sphere, the density of the sphere and of the liquid (measure these yourself, it's a nice filler), and also the terminal velocity of the sphere. You could devise a light gate system to record speed-time data as it accelerates from rest to its terminal velocity and deduce information on the force of friction. Or try a cylinder whose inside diameter is only slightly wider than the sphere and investigate what effect this has on the motion.

14   The Wind tunnel

Loved by some, hated by others; that's the wind tunnel. It's not a good choice for the timid student; but if you treat a setback as an opportunity to find out more, like to beat your own path through the world, or perhaps recognise the names Shipton and Tilman, this could be for you. It's got two potential stumbling blocks; the equipment and lack of variables.

Like any good bass speaker, you need to shift a lot of air. This takes muscle, so don't bother with hair dryers or tiny fans. The way to go is shown in the photograph at the bottom of the page. It's a 30cm diameter underground drainage pipe; corrugated on the outside, smooth on the inside, and 2m long. Just behind the grill sits a 12volt fan obtained from a car breakers yard. The gentleman responsible told me he had to search through many wrecks to find the correct fit for the pipe, and there's an obvious safety hazard (hence the grill), but the result is one very effective piece of gear which can be reused year after year. Air is drawn in through the open ended far side, and a clear plastic panel allows access to the experiment. The photograph also shows a vane type anemometer (currently about £120) for measuring the wind speed.

Measuring the lift force on a wing requires a small hole drilled in the bottom of the pipe. Pass a thin rod through this hole to connect the wing to a balance underneath the pipe (it's not as easy as it sounds, but if you've gotten this far, you'll manage it). For a given wing profile, measure the lift force as a function of the windspeed, and then try different wing profiles. Investigate the effect of scale; that is, what effect would it have on the lift force if the wing had a similar profile but was half the size? Try varying the angle of the wing, or photograph the air flow around the wing using white thread. Use the vane anemometer to measure the windspeed at different places across the cross section of the pipe.

The challenge is there, but so are the rewards.

15 Focal Length of Lenses

This is a good topic for a project. There are plenty of unfamiliar areas to explore, the equipment is readily available, and you already have an idea of the basics from Higher work. Its only drawback is the need for a darkroom, though a lab on the north side of the building with good blinds on the windows (pretty rare in my experience) will suffice. Many physics departments have a 'sixth year lab' where experiments can be set-up and left. They often have excellent blackout facilities and are ideal for optics experiments, but, it can get awkward when you need blackout and the other students need light. In my experience, the sense of camaraderie among Advanced Higher students always shows through. If available, an optical bench (long metal track for convenient mounting of object, lens and screen) is useful, but not essential. You'll find that cylindrical lenses are easier for measuring distances, but that there's more variety with spherical lenses (where measuring distances can be surprisingly tricky without the optical bench). Pay attention to safety regulations if you use a laser light source.

Start with the familiar thin convex lens and repeat the experiment to check the thin lens formula $\frac{1}{u}+\frac{1}{v}=\frac{1}{f}$.

The signs (+ve or -ve) of the quantities in optics are important, and just like HD DVD and Blue Ray, there are rival systems. Writing the formula as shown has already revealed the sign convention. It's called 'real is positive' and is used by Jenkins and White, authors of the greatest book on optics ever written. Do some research on signs. For a 20cm lens, position the object (a small 12V bulb) at regular distances on one side of the lens and measure the image distance. Draw a graph of the inverse of the object distance on the x-axis against the inverse of the image distance on the y-axis. The slope is -1 and the intercept (on both axes) is the inverse of the focal length. Repeat for lenses of different focal length, but don't spend too much time on it and view this section as the beginning of the project.

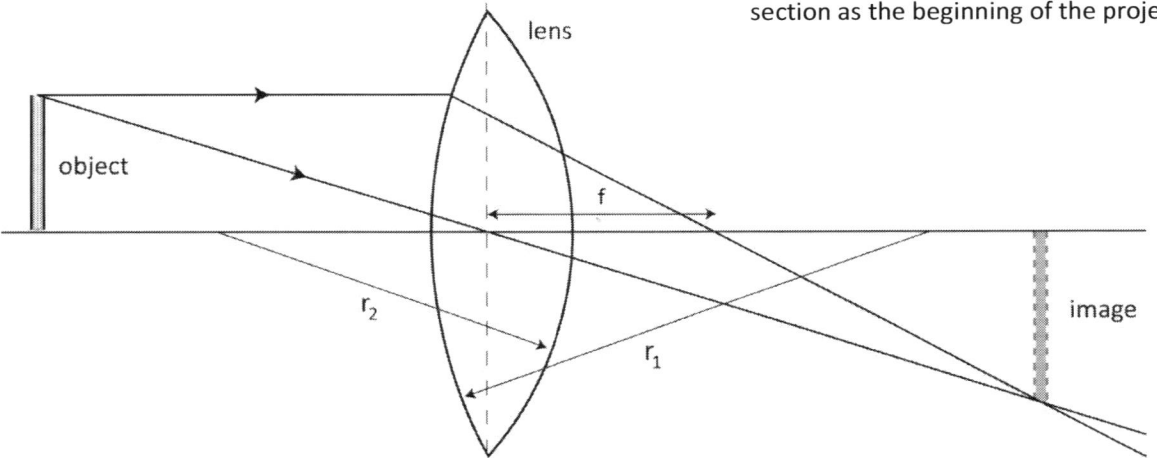

The focal length of a lens can be related to the curvature of its surfaces using the lens-makers formula $\frac{1}{f}=(n-1)\left(\frac{1}{r_1}-\frac{1}{r_2}\right)$. This version is for thin lenses (the diagram above is exaggerated a bit), where the thickness of the lens is much less than either radius of curvature. Most lenses have the same radius of curvature both sides (unlike the diagram), and using the correct signs, this is $r_1=-r_2$. You can try to measure the radius of curvature using a spherometer (like the one on page 26). What about the refractive index 'n'?

Having checked the thin lens formula, a productive area to explore is the imperfection of the image. A perfect lens should take the bright points on an object and map them as a faithful copy to form an undistorted image; but it doesn't. The two main imperfections ('aberrations') of the image are called spherical and chromatic. The surface of a sphere isn't the ideal shape for a lens; it's accurate near the centre but less so as you move outwards towards the edge. You can measure its effect on the distance to the focus point, by partly blocking off areas of the lens using black paper (what effect will this have on the brightness of the image for equal width gaps?) The diagram above shows the effect on a 20cm focal length lens using parallel light. Notice that the focus position doesn't vary all that much; a good test of your experimental skills. Investigating chromatic aberration requires different sources of monochromatic light (usually not practical except for sodium) or the use of coloured filters (readily available but they reduce the light intensity -especially the blue ones).

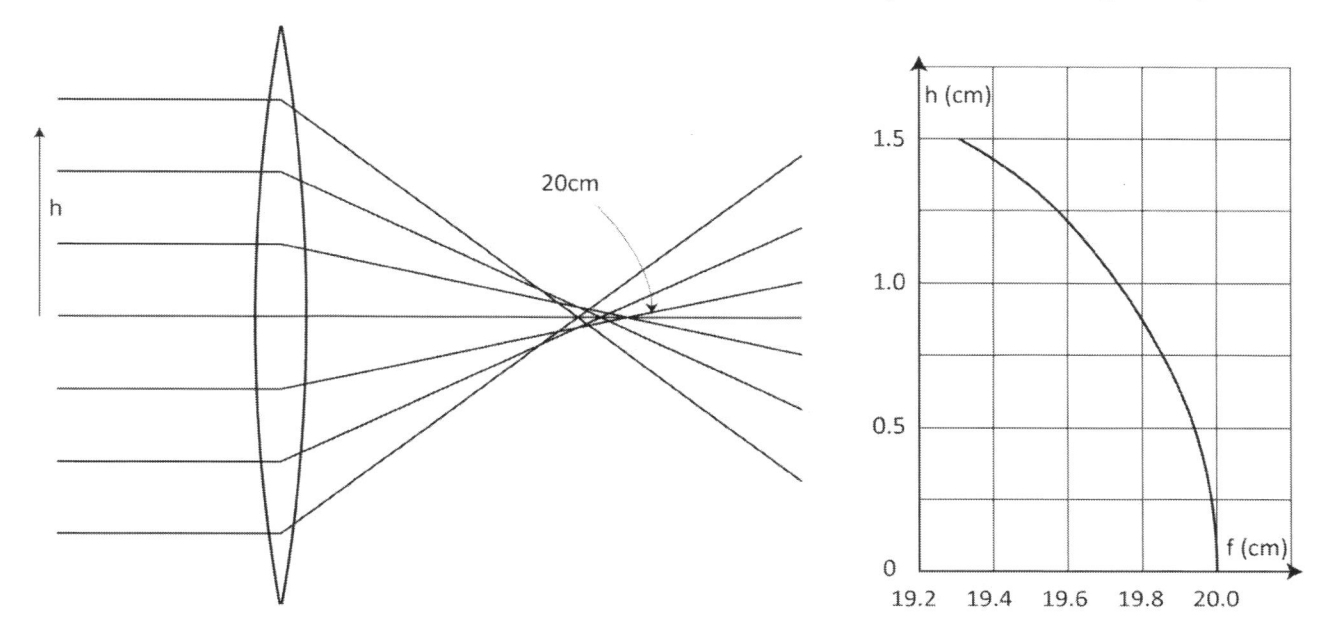

In a camera the image is detected on a flat 2D plane, regardless of whether it's using an old fashioned chemical film or a modern digital CMOS sensor. You can investigate what happens to the focus position when you move the object off-axis. Will the image come to a sharp focus on the flat image plane, or will it curve away from it? Move the small 12V light bulb in the direction 90° to the optical axis and plot the position of the focus point. Repeat for different object distances and for different focal length lenses.

If you've time for more experiments, measure the focal length of a concave lens using a pin, check if the combined power of two thin lenses in contact is the sum of the individual powers, or try compound lenses (combinations of single lenses which may be convex or concave). SLR cameras have detachable lenses. You could try to measure its focal length (if it's a zoom lens, like 28mm – 135mm, measure it at the extreme of its range). This is quite tricky. You could bring parallel light to a focus with it, but where would you measure from? For an interesting astronomical method using the constellation of Orion, try this website:

http://www.bobatkins.com/photography/technical/measuring_focal_length.html

16 The Helium-Neon Laser

Lasers are in just about every home in the country so it's hard to believe that it was once known as 'a solution looking for a problem'. It's a good topic for a project since you can start off in familiar territory, follow up with an interesting but straightforward measurement, and finish with a tricky bit. The three experiments I suggest are: measuring the wavelength as in Higher (but with more care and an uncertainty analysis), measuring the spread and intensity profile of the beam, and an attempt at measuring its power.

The standard method for measuring the wavelength of a monochromatic light source uses a diffraction grating. You either use the 'crossed metresticks' method, where you place the source (usually a bright sodium bulb) about 5m away, and look at it with the grating beside your eye, or use a laser source with the grating in front of it as shown below. For the latter method, check if its worthwhile taking care over the positioning of the grating and screen (do they have to be at exactly 90°?) How far away should be the screen? What number of lines per mm of the grating is best? Should you measure more than just the first order spots? Remember to note down the uncertainty in all your readings (what's the uncertainty in the line density of the grating?) Do a much more careful job than in Higher physics.

The equation for constructive interference of the myriad wavefronts from all the 'holes' is $n\lambda = d \sin \theta_n$. If you take readings for several orders, the SQA will expect you to extract the wavelength from the slope of a graph (rather than working each one out individually and taking an average). Think of it as an unexpected bonus. The equation becomes $\sin \theta_n = \left(\dfrac{\lambda}{d}\right) n$ and you plot the sine function on the y-axis against the order number 'n' on the x-axis. The slope will be the wavelength divided by the grating spacing. You should try to get within 10nm of the correct answer 632.8nm.

Ideally, laser beams would be parallel-sided, but in reality they spread out slightly. The second experiment quantifies this angular spread and picks up more bonus points by measuring the intensity profile across the beam using a photodiode. The beam is at its most intense at the centre, then rapidly drops off towards the edge. How you define the width of it is up to you, but between the points where the intensity halves would be a reasonable choice. This is why drawing the profile is important, and to do this requires a small photodiode like the one on the right. It's from the Osram BPX series and is very common.

The photodiode looks about 1mm across but the datasheet gives the sensitive area as only 0.675mm^2. Use this to decide how much you can advance the photodiode for the next reading across the beam. Don't hold the photodiode with a clampstand; it's totally inadequate. The photograph above shows a much better way of moving the photodiode accurately over a small distance. Notice the fine threaded bolt and the spring. The bolt moves through a snug-fitting nut (embedded in the wood with araldite) to prevent backlash. The square sectioned piece of wood heading out of the picture on the right is clamped firmly to the bench for overall stability.

The photostrip on the right shows (from top to bottom) the beam broadening as the laser is moved farther back. The distances were 1.5m, 3.6m, 5.6m and 7.75m and the beam is shown above the photodiode to make the illustration clearer. Take as many readings as you can and plot several profiles. Decide what definition to adopt with the width of the beam and use simple geometry to calculate the angular spread of the beam (called the 'divergence', in milliradians). Remember to do the uncertainties.

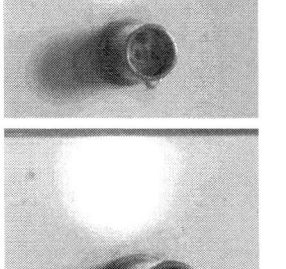

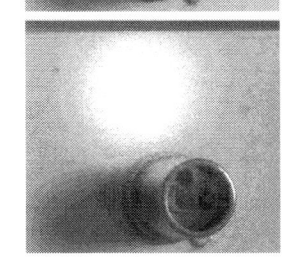

The third experiment is straightforward in principle but can be surprisingly tricky. It's for the student who likes a challenge and has the requisite degree of persistence. The idea is to measure the power of the laser by shining the beam on a surface and measuring the rise in temperature. This is how commercial companies do it using devices like the one pictured below (the sensor is in the rectangular housing and it connects to a digital meter calibrated to display the power). The method sounds simple. Your laser hasn't got much power though (it wouldn't be in school if it did), so you must shine it on a small object and expect a small temperature rise. An obvious object to try is the thermistor; it's sensitive to small changes in temperature, and they also come in small sizes. The one in the photograph above is about 8mm in diameter and 1mm thick. Your first exercise is to discover/measure its mass (your teacher might not let you snip off the wires, so that could be your first hurdle). The second one is to find out / discover its specific heat capacity (you might need educated guesses here). With these two numbers, you then perform a 'pre-calculation'. What temperature rise (or the rate at which the temperature should rise) would you expect assuming the laser had the maximum power of 1mW allowed under the class 2 designation? Using energy conservation (all the light energy is absorbed) gives:

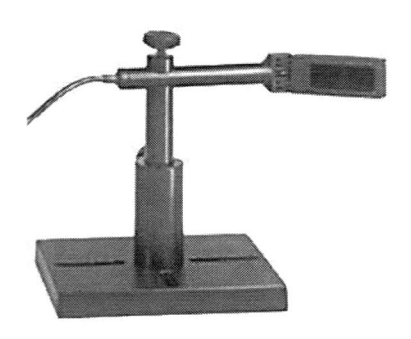

$$Pt = cm\Delta T \quad \Rightarrow \quad \frac{\Delta T}{t} = \frac{P}{cm} = \frac{0.001}{700 \times 0.0001} \approx 0.014 \,°C/s$$

I've pretty well guessed the numbers but it gives you a rise of about 1°C every minute, and that's for the maximum assumed power. Having calibrated the thermistor using hot/cold water and an accurate thermometer, you'll know what to expect for the change in resistance of the thermistor.

What practical problems will you encounter? The main one is thermally isolating the system from anything except the laser. You must give the system plenty of time to reach thermal equilibrium before starting, so have it set up hours beforehand in a box of your own design (and consider that when you start the laser, the thermistor will heat up non-uniformly). A good multimeter reading to 0.01Ω will help. Plotting a temperature / time graph will let you spot any trends. As I said at the start, it's a great project for a certain kind of student. It pushes at the limits of what you can achieve. Best of luck!

Laser safety

The laser you'll be using is called 'class 2'. This type emits visible light and has an output power of less than 1mW. You must carry out a risk assessment and be aware of your responsibilities. There are obvious things like 'don't look into the laser', but you can be caught out with reflections from mirrors etc. It's sensible if you get everything set-up first, only switch on the beam when taking readings, keep the beam parallel to the bench, and not leave it unattended. Don't go off for lunch with the key still in the lock.

17   Musical Sounds

Music is one of those rare instances where science meets the arts. A student, who can play an instrument such as a guitar or violin, could find some interesting connections between them with this project and enhance their appreciation of both. What you must avoid is being sucked into the gaping maw of IT; clicking mouse buttons all day long isn't physics. Use any software as a tool, but don't let it become the project.

You can get the physics off to a good start with stringed instruments. A wire vibrating by itself will only make a quiet sound since it doesn't push against much air. But make it transmit its vibrations to a sounding box and the box will shift a lot more air to give a louder sound. Most schools will have a sonometer like the one in the photograph. Set up the way it is, it looks like the standing wave experiment without the vibration generator. That experiment was an example of forced vibrations where you invite the wire to vibrate at any frequency, whereas this one will vibrate at the frequency of its own choosing, the 'natural frequency'. Since the wire is held down at the knife edges (making them nodal points), the lowest frequency standing wave will be half a wavelength. The harmonics will be the frequencies which fit in more waves along the wire but retain nodes at the knife edges. If you pluck the wire, the actual frequency generated depends on the tension in the wire and the mass of the wire (or more correctly, the mass per metre length of the wire). The equation you'll be using relates these quantities to the wavespeed. For the natural frequency, the wavelength is fixed (equal to twice the distance between the knife edges), so the wavespeed is proportional to the frequency you hear (from $v = f\lambda$).

The expression is $v = \sqrt{\dfrac{T}{\mu}}$ and you should give a well laid out derivation of it for your Report. The 'T' is the tension of the wire in Newtons (pretty-well the same as the weight hanging from the end of the wire if there was no friction at the knife edge), and the 'μ' is the mass per metre of wire in kg m^{-1}. The designer of a stringed instrument with a fixed length of wire only has these two variables to play with in determining the available range of frequencies. This is why a guitar has light strings under high tension for high notes, and heavy, wound strings under low tension for the bass notes (and also why frets had to be invented). The photograph on the next page shows the metal strings on a six string guitar. For the scale, the distance between top and bottom strings is 4.6cm. I've included the derivation of the wave equation beside the photo so you can understand where the expression for the wavespeed comes from. Go through it line by line, fill in the missing steps, and try to understand it.

This diagram shows a short section ΔL of the wire. I've exaggerated the curve to make things clearer.

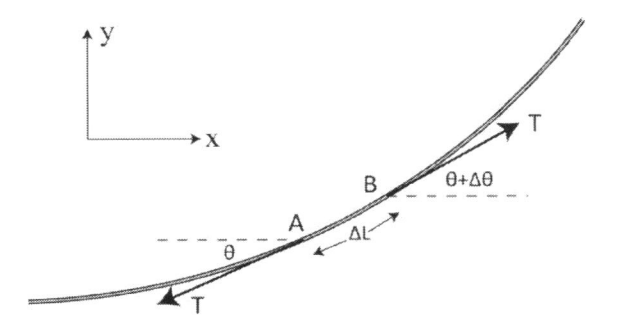

For small displacements of the string, the tension 'T' at any point of the string stays about constant in magnitude. The 'y' direction is 'up and down'. Taking the 'y' component of the tension at 'A' (acting down) and the 'y' component at 'B' (acting up), gives a resultant force on the section of length ΔL in the 'y' direction as:

$$\Delta F = T \sin(\theta + \Delta\theta) - T \sin\theta$$

The mass of length ΔL is μΔL, so Newton's 2nd Law gives:

$$ma = \mu \Delta L \frac{d^2 y}{dt^2} = +T \sin(\theta + \Delta\theta) - T \sin\theta$$
$$= +T \Delta\theta \cos\theta$$

(the second step takes a few lines of trigonometry). Now work on the left side. Use the tangent at a point:

$$\frac{dy}{dx} = \tan\theta \quad \Rightarrow \quad \frac{d^2 y}{dx^2} = \sec^2\theta \frac{\Delta\theta}{\Delta x} \qquad \text{Then use:}$$

$$\Delta x = \cos\theta \Delta L \quad \Rightarrow \quad \frac{d^2 y}{dx^2} = \sec^2\theta \frac{\Delta\theta}{\Delta x} = \sec^2\theta \frac{\Delta\theta}{\cos\theta \Delta L}$$

Bring together and tidy up:

$$\mu \frac{d^2 y}{dt^2} = +T \cos^4\theta \frac{d^2 y}{dx^2}$$

For small amplitudes, the cosine part is about one, so

$$\mu \frac{d^2 y}{dt^2} = +T \frac{d^2 y}{dx^2} \quad \Rightarrow \quad \frac{d^2 y}{dx^2} = \frac{\mu}{T} \frac{d^2 y}{dt^2}$$

Check that both $y = A\sin(x - vt)$ and $y = A\sin(x + vt)$ are solutions of this equation with the expected expression for the speed.

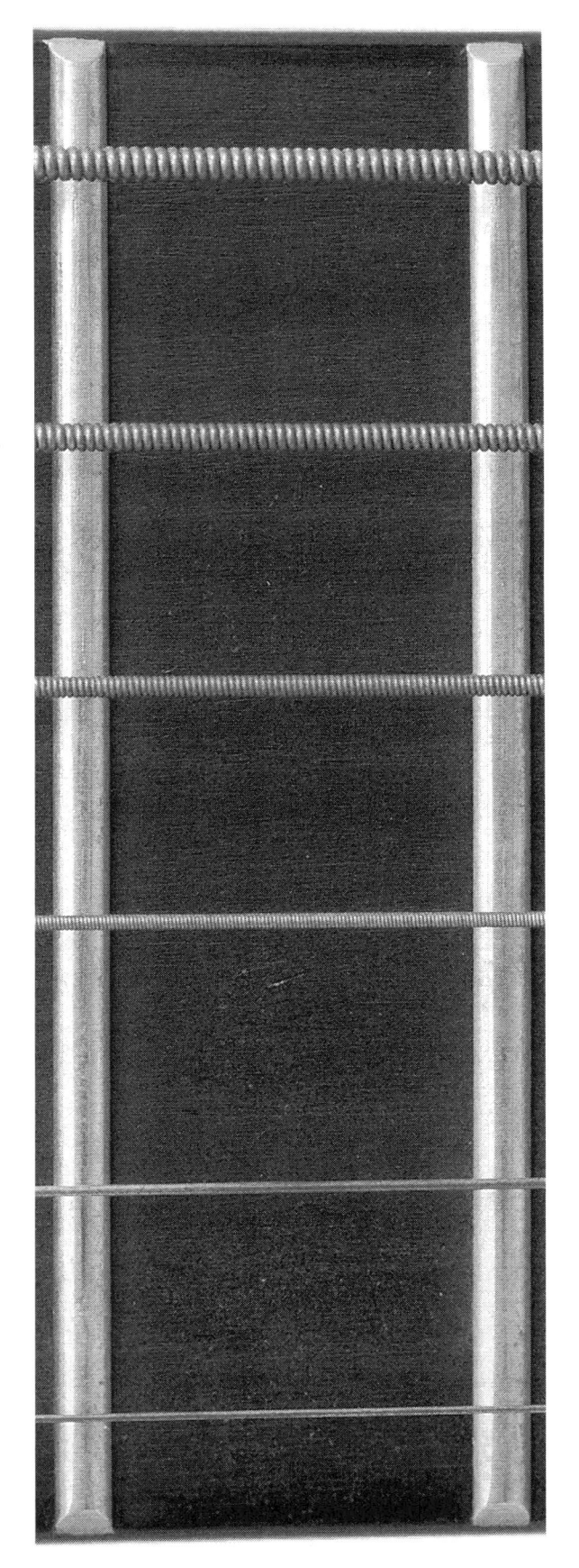

Using $v = f\lambda$, and $\lambda = 2L$ for the fundamental, we get $f^2 = \dfrac{T}{4\mu L^2}$. Your experiment is to show that $f^2 \propto T$, $f^2 \propto \dfrac{1}{\mu}$ and $f \propto \dfrac{1}{L}$. The first requires different weights and the second requires different wires. Do an accurate determination of the mass per unit length for each one (remembering uncertainties). The question is how to measure the frequency of the natural vibration. The easiest way is to set up a strobe light beside the wire, adjust the strobe frequency to the lowest frequency for single viewing, then read the dial. A traditional method uses tuning forks and listens for a beat frequency. You place the pointy base of the vibrating tuning fork on the knife edge at the weight end and adjust the position of the knife edge until the wire has the same frequency (shown by the beats disappearing). Notice the displacement of the wire. Another method of measuring frequency would pass a small current through the wire (croc clips outside the knife edges) and generate a voltage in a coil beside an antinode. Feed the output to an oscilloscope or an audio amplifier and frequency meter. A fourth possibility is to detect the sound with a microphone, connect to the soundcard of a computer, and analyse the results at your leisure (you might also see signs of the overtones produced by the sounding board). The above experiments with the wire could be the main physics element of your report, so make sure and do a comprehensive job of it.

The second section of the investigation could analyse the sound from real instruments. This used to be done using either an exotic object like a storage oscilloscope, or taking photographs of waveforms on a standard oscilloscope. The widespread use of computers has really made these methods redundant (yes, it's not as good for your brain, but how many teachers still use ticker timers for acceleration experiments?) A useful (and FREE) piece of software is called Audacity. Your school probably uses it already (if they don't, then it's a no strings attached download). Take some time learning the features on the toolbar shown below. Of particular interest is the small 'magnify' icon fourth from the right. This pulls out the waveform just like the timebase on an oscilloscope. Within the 'analyse' pull down you'll find 'plot spectrum', which allows you to identify the frequencies making up your signal. Experiment with it. The two shown below of a bass guitar and piano are both relatively clean signals taken from .wav samples off cds. Using 'plot spectrum' identified the fundamental frequency of the bass note as 73Hz. When you're ready, record your own samples of several seconds duration

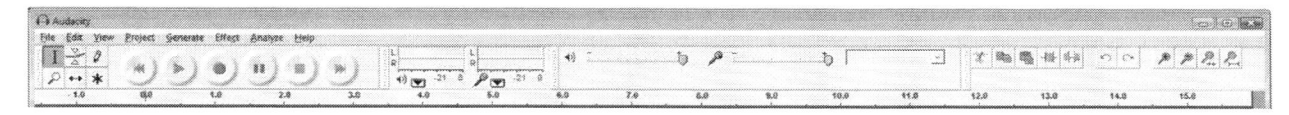

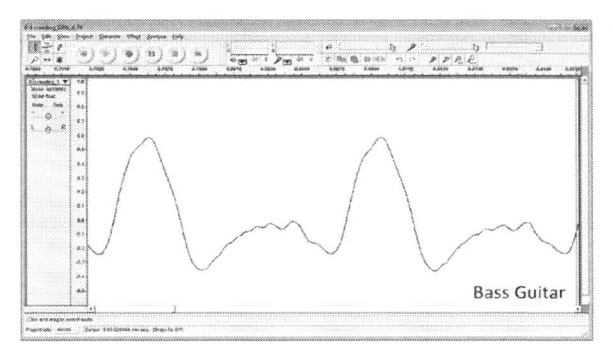

Bass Guitar

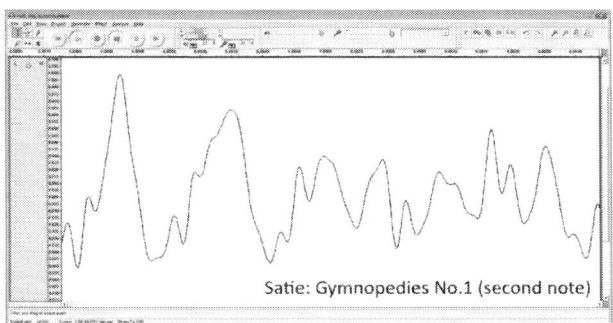

Satie: Gymnopedies No.1 (second note)

using real instruments. Take samples of different pitches for string, wood and brass instruments if you can get access to them (might need the help of a musical friend). It's always best recording your own samples with a microphone rather than downloading them from the internet.

As I've been stressing, the danger is that you end up making things too 'computer' and not enough 'physics'. One area to investigate is to take a guitar string and pluck it at different positions. Here is an example showing the difference between a guitar string plucked at the middle of the string (left side illustrations, showing the

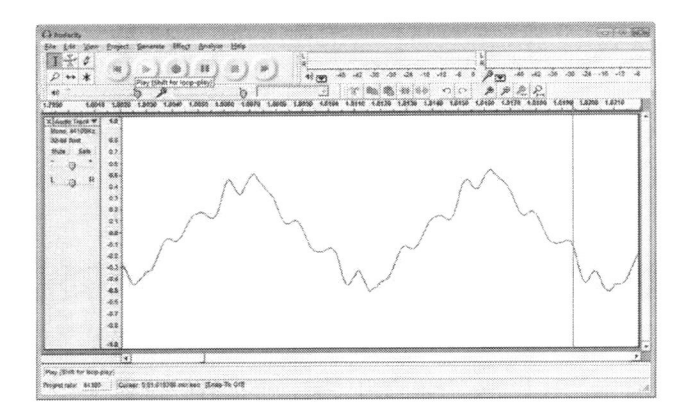

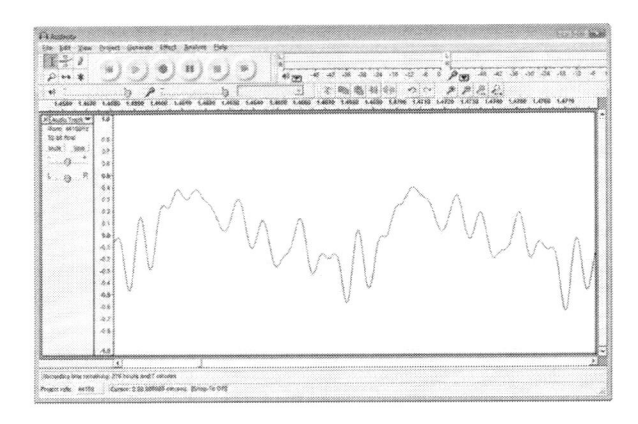

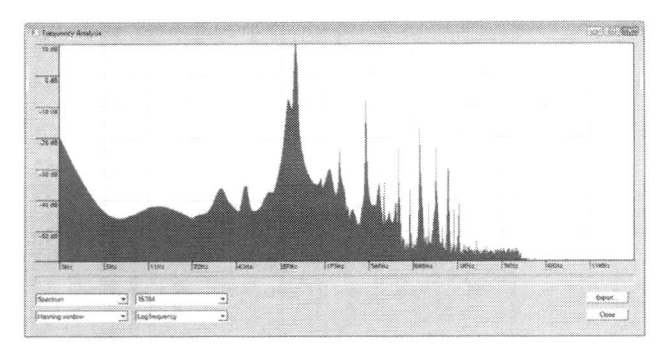

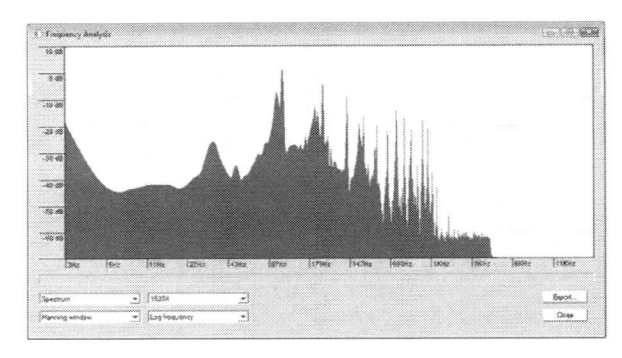

waveform and frequency analysis) and plucked near the bridge. They are (and sound) quite distinct, and the difference is due to the response of the sounding board. The fundamental frequency is the same in both cases, but the overtones have different weights. A basic web demo of the effect of adding overtones is at: http://www.bsharp.org/physics/stuff/Overtones.html You can choose which overtones to add, but not the weight. Have Audacity loaded, press the 'play loop' button on the web page, then press the 'record' button on Audacity at the appropriate moment. Stretch out with the 'magnify' button.

An interesting area for those with more time, is the study of vibrations of 2D surfaces like a drum skin or a bell. The overtones have no simple relation to the fundamental frequency, which is why a drum doesn't sound 'musical'. And yet a similar surface (from the physics perspective!) like a bell, does sound musical. An investigation of the overtones recorded from real bells (and what makes them musical) could be a chance at some real research. Look up 'Chladni's figures' on the internet.

18   The Bike Wheel

The ordinary push-bike is a source of lots of good physics and should suit the student who's a keen biker. It uses the material recently learned on rotational motion, presents opportunities for fiddling about with spanners, and generates lots of data. Add in the need for a knowledge of computer interfacing and familiarity with Excel, and you have the basis for an excellent project. There are many possibilities for an investigation but I suggest studying the motion of the front wheel. It'll be more than enough for the Report. Split it into two parts: the study of a flywheel for the first part, followed by the bike wheel.

Most schools will have a flywheel. Some of them will be heavy enough to damage your toes. It's a device for storing rotational kinetic energy and it does this by having a large moment of inertia. In this respect it differs from the bike wheel where the object is to minimise the moment of inertia (think what it would feel like if you didn't). But they both share the same physics and the parameters of the flywheel are easy to handle since it has a simple shape. Hopefully yours will consist of a combination of cylinders like the one shown below. You can calculate the moment of inertia if you measure the dimensions and mass of each bit. The object of the first part of the project is to compare your theoretical prediction with the experimentally measured value. For a flywheel consisting of cylinders, use $I = \frac{1}{2}mr^2$ for each part (having solved the problem of not being able to dismantle it, by using the density).

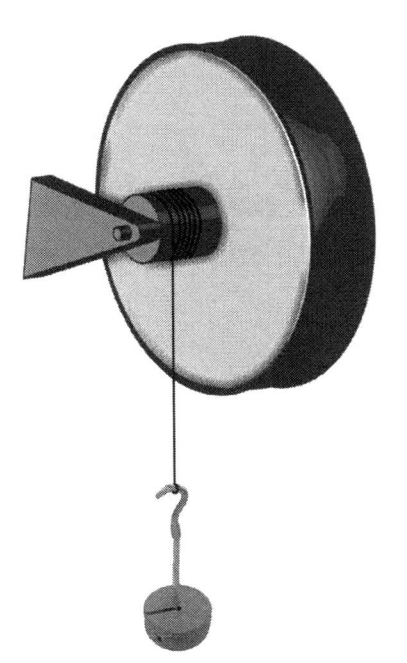

To measure the moment of inertia, firmly fix the flywheel about one to two meters above the floor. Attach the string so that it doesn't slip, but falls off the spindle when the weight reaches the floor (traditionally it's a small peg on the spindle). Count the number of rotations 'N_1' that the flywheel will make by backing up the weight from the floor to the start position. Twenty turns would be an angular displacement of 40π radians. Time how long it takes the weight to reach the floor and use the results to calculate the average, then final 'ω', angular velocity. Measure the starting height 'h'.

By conservation of energy:

$$mgh = \frac{1}{2}I\omega^2 + \frac{1}{2}mv^2 + E_{Friction}$$

The energy 'lost' to friction will derive from the bearings. A common assumption is that the energy generated due to frictional forces doesn't depend upon the speed. The sound and heat energy generated for one complete turn will be the same at the start and the end of the motion. Suppose the energy lost per turn is 'E', then for 'N_1' turns, the total energy generated due to friction is N_1E. The equation becomes:

$$mgh = \frac{1}{2}I\omega^2 + \frac{1}{2}mv^2 + N_1E$$

Once the string slips off the spindle, the only force acting will be due to friction. The final rotational kinetic energy of the flywheel will be converted into sound and heat energy due to friction, and bring it to a stop. Count the number of turns 'N_2' from when the weight reaches the floor until the flywheel comes to a stop. Using energy conservation for this part, and with the same assumptions as before:

$$\frac{1}{2}I\omega^2 = N_2E$$

Eliminate the 'E' from these two equations to get:

$$mgh = \frac{1}{2}I\omega^2\left[1+\frac{N_1}{N_2}\right]+\frac{1}{2}mv^2$$

The final speed of the weight 'v', is given by $v = \omega r$ where 'r' is the radius of the spindle. Substitution gives:

$$mgh = \frac{1}{2}I\omega^2\left[1+\frac{N_1}{N_2}\right]+\frac{1}{2}m\omega^2 r^2$$

You've now measured all the quantities in the equation needed to calculate the moment of inertia. Using the same starting height, repeat the experiment and check if the number of turns 'N_2', is the same as last time. Then repeat for different starting heights. Hopefully, the moment of inertia will be about the same each time.

Move on to the bike wheel for the second part of the Investigation. The picture below shows the two main designs. The mountain bike arrangement on the left has more mass in the tyres (though not the rims) and a more complicated hub. It also has a brake disc. The road bike has low mass tyres and a simple hub shape. For a calculation of moment of inertia, the road bike would be more straightforward, but if the object is simply to *measure* the moment of inertia, then either design is suitable.

Repeat the first experiment using a bike wheel rather than the flywheel. If practical, measure the masses and dimensions of the components and calculate the moment of inertia. Take the wheel to be a hoop (for the rim, or two hoops if you include the tyre) spun about its centre, a number of rods (for the spokes) spun about one end, and a cylinder (for the hub) spun about its centre. You have to decide whether to leave on the tyre (making it more realistic), or remove it. Notice that the spokes aren't pointing straight at the centre of the hub. They are connected at (or nearly at) a tangent to the outer hub rim. It's not just the chain that transmits the force of your muscles to the road, it also goes through the spokes, and a tangential placement is more efficient (possibly scope for some work here using torques). If you calculate the three contributions to the moment of inertia, consider the relative size of each one and where to put in the effort in calculating the uncertainties.

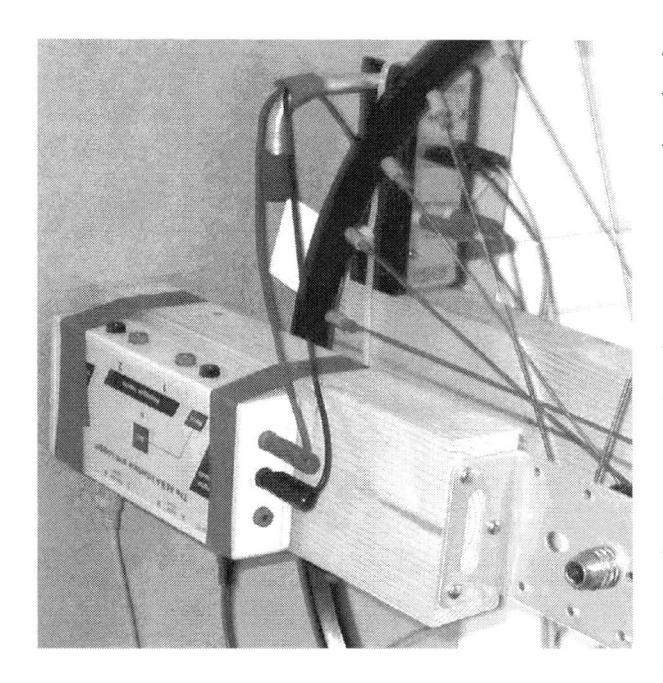

An alternative method to the one above is to measure the angular velocity as a function of time. With a tyre on the rim, the resistance due to friction comes from the air resistance of the tyre as well as the frictional force at the hub bearings. If you remove the tyre, the air resistance decreases and the physics is like that of the flywheel. The flywheel calculation assumed that the force of friction was independent of the speed of the wheel; but the force against the motion due to the air resistance of the tyre will vary as a power of the speed of the tyre. Monitoring the angular velocity against time as the wheel slows down due to friction lets you to analyse the forces. The arrangement on the photograph (without a tyre!) uses a light gate where the beam is cut by the spoke nipples. If the recorded times are too short, or the nipples aren't circular, slip rubber tubing over them as shown. Rotate the wheel by hand and allow it to slow down. Send the data (with 32 spokes and several turns, that's a lot of numbers) from the interface to your laptop where you have Excel waiting. Input the diameter of the nipple and its distance from the centre of the hub, then use Excel to calculate the angular velocity and the total angular displacement (from a count of the spokes). The next two pages show sample results; the first one using the falling weight showing the angular acceleration of the wheel, and the second one with the wheel coming to rest after an initial push by hand.

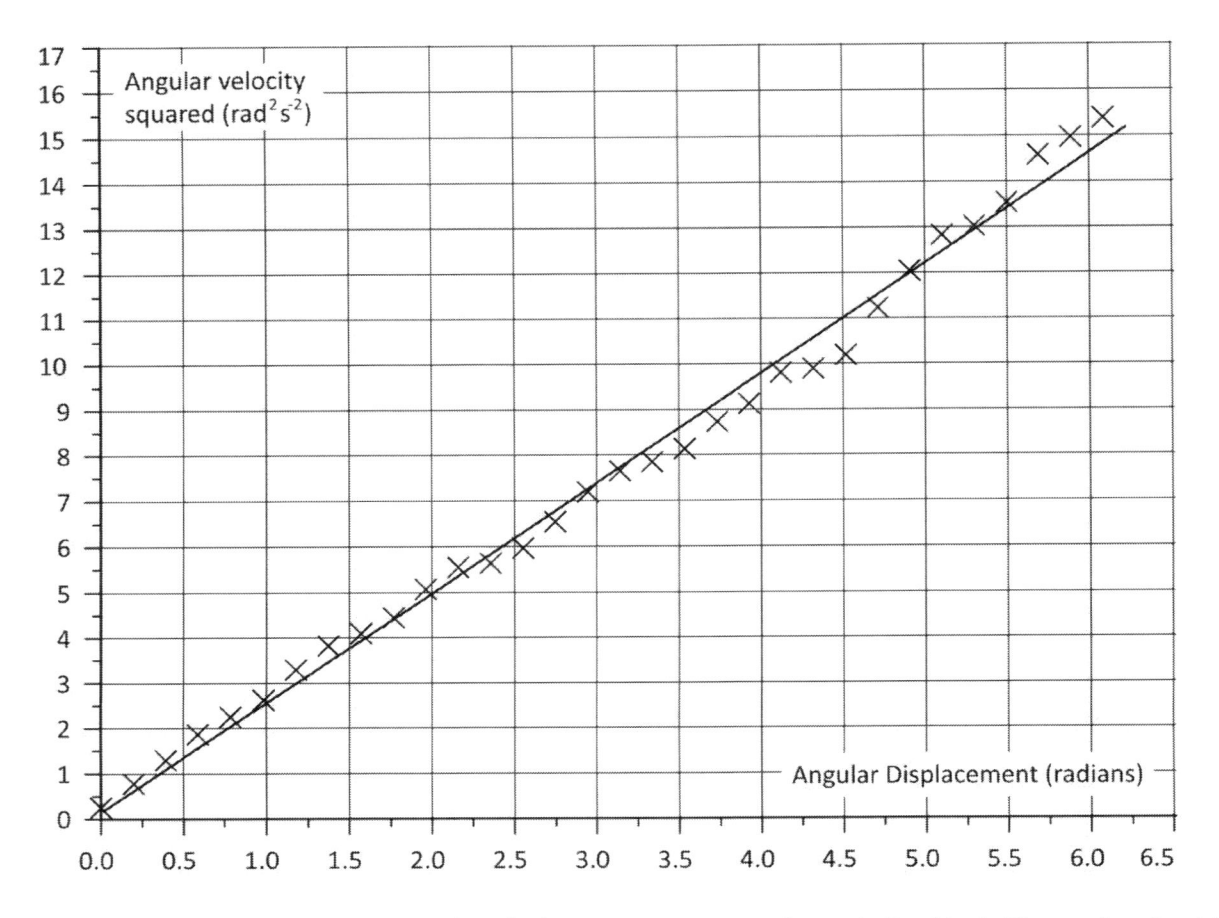

A good experimenter should have some idea of what to expect, yet not be prejudiced by it. The motion should be close to one of constant angular acceleration where $\omega^2 = \omega_o^2 + 2\alpha\theta$, so we plot the squared angular velocity against the angular displacement. The intercept will inform us of the initial angular velocity. You would think it would be zero, but it's when the angular displacement starts from zero, and that's when the first spoke cuts the light beam – you may have released the weight with the light beam anywhere between two spokes. We can deduce the angular acceleration from the slope.

The graph shows a best-fit straight line derived from Excel 2003. Its equation is:

$$\omega^2 = 0.19 + 2.4\theta = 0.44^2 + 2 \times 1.2\theta$$

Giving an initial angular velocity of 0.44 rad s^{-1} and angular acceleration of 1.2 rad s^{-2}. To work out the moment of inertia of the wheel, we need to calculate the resultant torque:

$$T_{app} - T_{frict} = I\alpha \quad \Rightarrow \quad T_{app} = I\alpha + T_{frict}$$

The torque applied by using the weight $T_{app} = rF$ is due to the tension 'F' in the string (it's not the same as the weight $W = mg$ though it will be roughly correct since the masses used will be a lot less than the wheel – see the Theory Book page 36). Repeat the experiment several times using different masses (typically from 10g up to 80g), and measure the angular acceleration for each one from a graph as above. Plot the applied torque on the y-axis and the angular acceleration on the x-axis. The gradient will be the moment of inertia and the intercept will be the frictional torque. Remember to do the uncertainties.

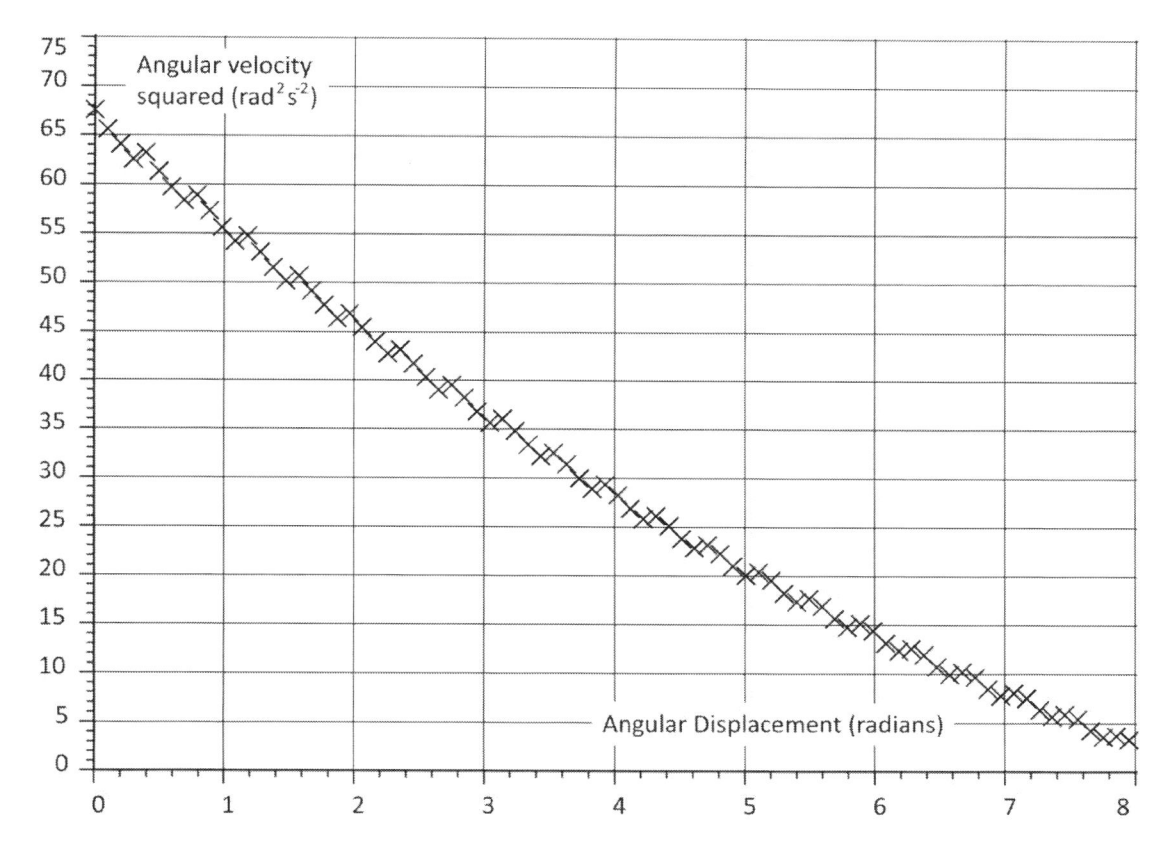

Excel 2003 has a 'Trendline' panel for fitting functions to curves like the slowing-down dataset above. Use your knowledge of the physics of the situation to choose the best function. The behaviour of the graph at the start has a definite intercept, so there's no point trying to fit a power law like inverse proportionality. When it slows down, the graph line approaches the x-axis like an asymptote, but must come to a stop. If you try a polynomial fit (second order) to the above curve, you get:

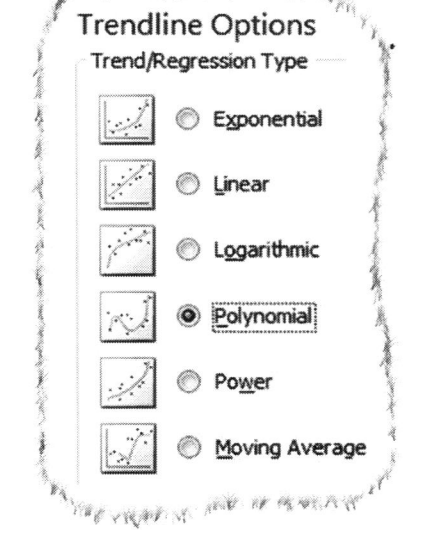

$$\omega^2 = 66.7 - 11.4\theta + 0.44\theta^2$$

This fits the graph well, but it's too phenomenological. Differentiate both sides with respect to time, cancel the angular velocity term, and you'll have an expression for the angular acceleration as a constant part and an angle-dependent part. Try to interpret this. An exponential decay curve might give a better fit. The Excel version gives this result:

$$\omega^2 = 67.6 e^{-0.28\theta}$$

Unfortunately, it's a poor fit (it's too concave). So what can you do? The slope at a point is twice the angular acceleration, so you could measure the slope by hand at the beginning and at the end. With no driving torque, the equation of motion is $T_{frict} = -I\alpha$. For the end of the motion, check if your previous results for the moment of inertia and frictional torque for the driven wheel are consistent (it should be in this case since it only went up to $\omega^2 = 16$). Students who like a problem should try to reach conclusions regarding the frictional torque due to the air resistance component!

19 Filament Light Bulbs

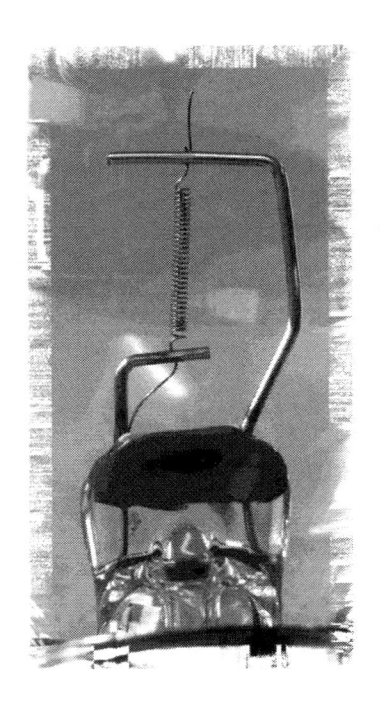

The filament light bulb is out of favour in todays' environmentally conscious world. Most of its energy is emitted in a form that your skin can feel but your eyes can't detect. That makes it unsuitable for street lighting (unless you're a bird), but in fact it's not such a bad thing in a house. Some day, houses will become super-insulated, and the heat energy from filament light bulbs and their owners could provide comfortable living spaces.

In the meantime, this much derided little bit of curly wire will give you plenty of material for a physics Investigation. The one on the right is an SBC 12volt 24watt raybox bulb commonly found in schools. Start by repeating an experiment you will remember from Standard Grade physics – just measure the voltage and current, then plot a graph! This time, you will include uncertainties, and do a proper analysis of the results. Drive the filament a bit harder than usual with two dc power supplies in series (connect them just like two cells). The object is to worry the bulb, but not to break it (take it up to about 14volts on the voltmeter). The graph below shows you what to expect. The gradient of the curve is the resistance, and increases as the temperature of the filament increases.

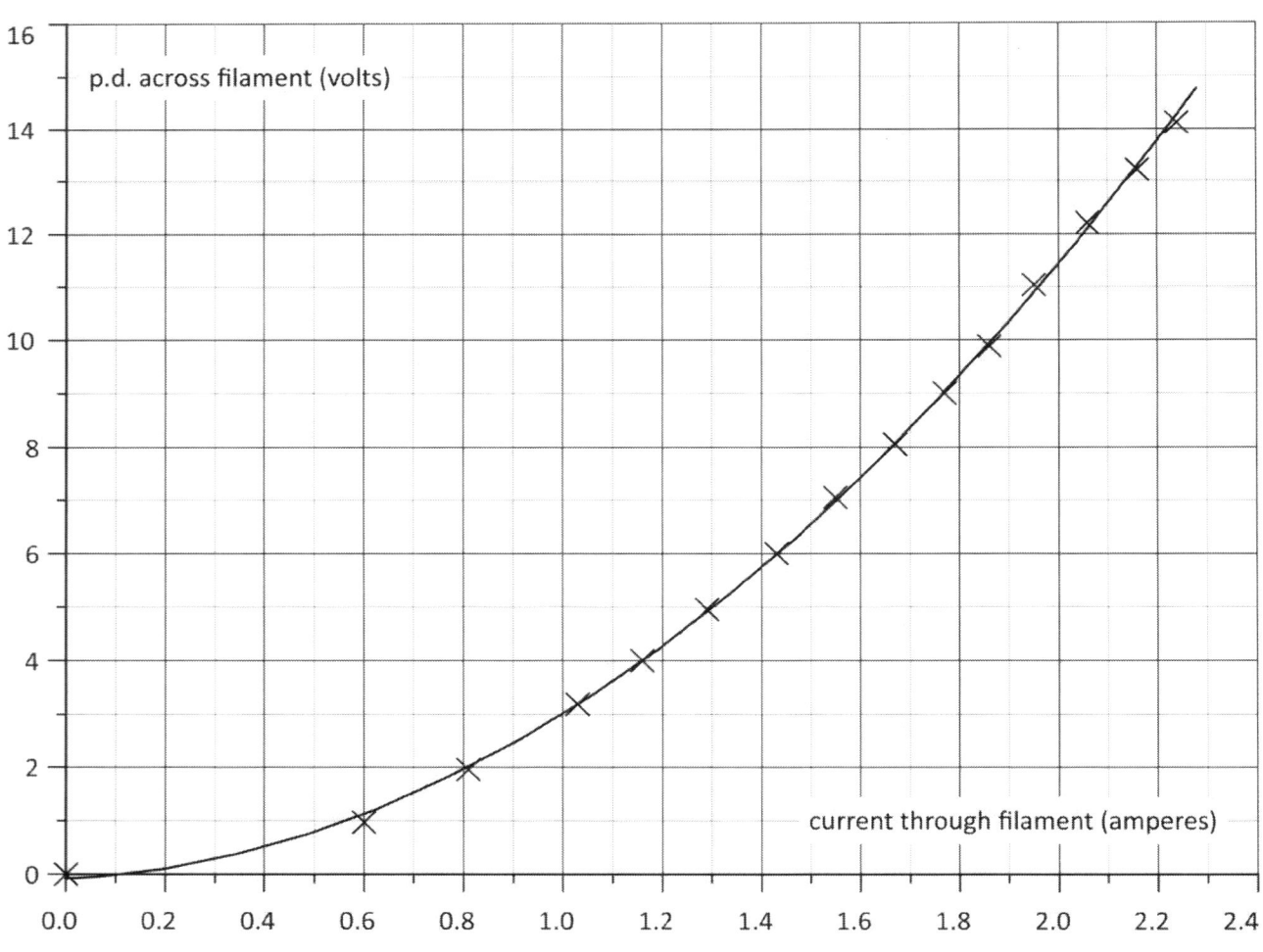

Use Excel to fit a second order polynomial to the curve. For that particular curve, I got:

$$V = -0.086 + 0.426I + 2.667I^2$$

Now calculate the resistance and power as a function of the current:

$$R = \text{(slope of curve)} = \frac{dV}{dI} = 0.426 + 5.33I \qquad P = VI = -0.086I + 0.426I^2 + 2.667I^3$$

At what current would the filament achieve its rated power output of 24watts. What would be its resistance?

The next stage is a bit trickier and tries to measure the surface area of the filament. This is an important quantity. Although the energy is generated by the volume of the filament, the surface area limits how much is radiated. Below is a photograph of the filament used in the last experiment, sitting on its side. It's a single coil

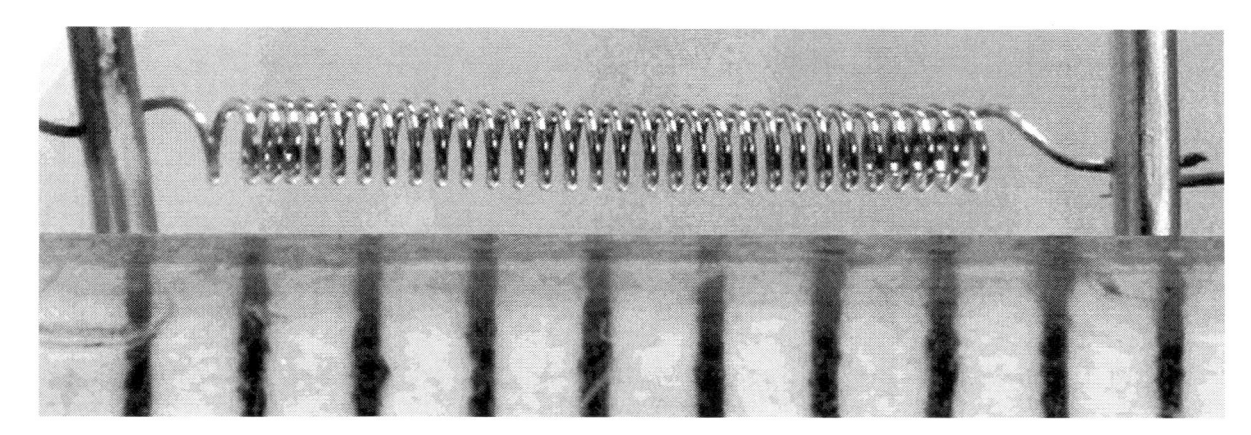

rather than the 'coils of coils' type (a design which increases the area), so it's a bit simpler to work with. A millimetre scale is beside it, and you could use that to calculate the area. Treat the filament as a long cylinder, work out its length then multiply it by the circumference of the wire. That last part is very small and has a high uncertainty. If you can arrange to take a close-up photograph of your filament, then have a go at it (try the Art department; they often have keen photographers). The one in the photograph was taken without breaking the glass using a narrow depth of field aperture on the camera. It was replaced by a ruler at roughly the same distance which was then moved until the ruler was back into sharp focus. This ensures the scale is accurate when the two images are superimposed in Photoshop. If you are allowed to break the glass, then use clamp stands to hold the bulb and the ruler underneath it. The bottom photograph shows the other common 12volt 24watt raybox bulb supplied by RS Components. This one is smaller and very tricky to measure.

If the measurements prove too difficult then try this alternative method. It uses Stefan's law $P = \sigma T^4$ which relates the power output of the filament to its temperature. The constant 'σ' is 5.67 x 10^{-8} W m^{-2} K^{-4}, and gives you the power output for an area of one square metre of filament (hence the m^{-2} in the unit). The equation is accurate for ideal radiators called 'black bodies', and at high temperatures the filament should be a reasonable approximation. You can easily measure the power output for any setting using the previous work, and you can estimate the temperature from the colour of the glowing filament. As you increase the current, the filament goes from a dull red, to an orangey colour, then towards white and a pale blue-white. Place the enclosed coloured card behind the filament. Start off with a low current and match up the reds and oranges with the temperature along the bottom. Try it for several different power outputs; you should get the same answer for the surface area of the filament (even better to plot the power up the y-axis and the fourth power of the temperature in (Kelvins)^4 along the x-axis; the gradient will be Stefan's constant multiplied by the area of your filament). When discussing this method, consider the effect of the fourth power of the temperature on the uncertainty. You'll have noticed that the ends of the filament are a bit cooler since they are conducting heat energy through to the pillars (use the centre part for readings). It's possible to draw fairly rough temperature profiles along the length of the filament for different low settings of the current (distance on the x-axis and temperature up the y-axis).

Another possible alternative uses the density of the filament. Use your photograph only to estimate the length of the filament. Cut the filament at its ends and record its mass on a balance. Look up the density of tungsten from a datasheet and calculate its volume. Calculate its cross sectional area from the volume and length, and hence its surface area. Check Stefan's Law using previous (pre-cutting) data for the power and temperature using the colour chart. The problem? Your school needs to possess a very expensive balance to get a reading!

The third experiment maps the distribution of light emitted by the filament. The filament shown at the start of this topic is mounted vertically (like the one in a raybox where it lines up with the slits). Circling around it in a horizontal plane at a fixed distance would give a constant light level. Other filaments are mounted horizontally. The one on the left is a 12volt 21watt bulb supplied by RS Components, and on the right is a 12volt 50watt halogen bulb from a well known Scandinavian retailer. The distribution of light in the horizontal plane would not possess circular symmetry. Viewed end-on, you would receive less light. Set up a simple circuit using the usual reverse biaised photodiode and record the current reading at various angles around the horizontal plane centred on the filament. Try different bulbs.

You could also try a normal mains light bulb of the 'clear' glass type (make sure the wiring arrangement is safe before you switch on – no bare wires showing). This photograph is the common half hexagon arrangement of many mains bulbs. It is of the 'coils of coils' type to increase the radiating surface area. You could calculate its surface area using Stefan's Law as before. Placed horizontally it should give a reasonably uniform light intensity. Using a photodiode, readings for the graph below were taken every 15° at a constant distance of 10cm from the filament. That graph is called a polar plot. It's one way of displaying your results. You could also show them using a two dimensional area centred on the filament and plot the intensity at various distances (with your page covered in numbers). More adventurous still, would be a three dimensional plot, and by this time, you're well on the way to lighting design.

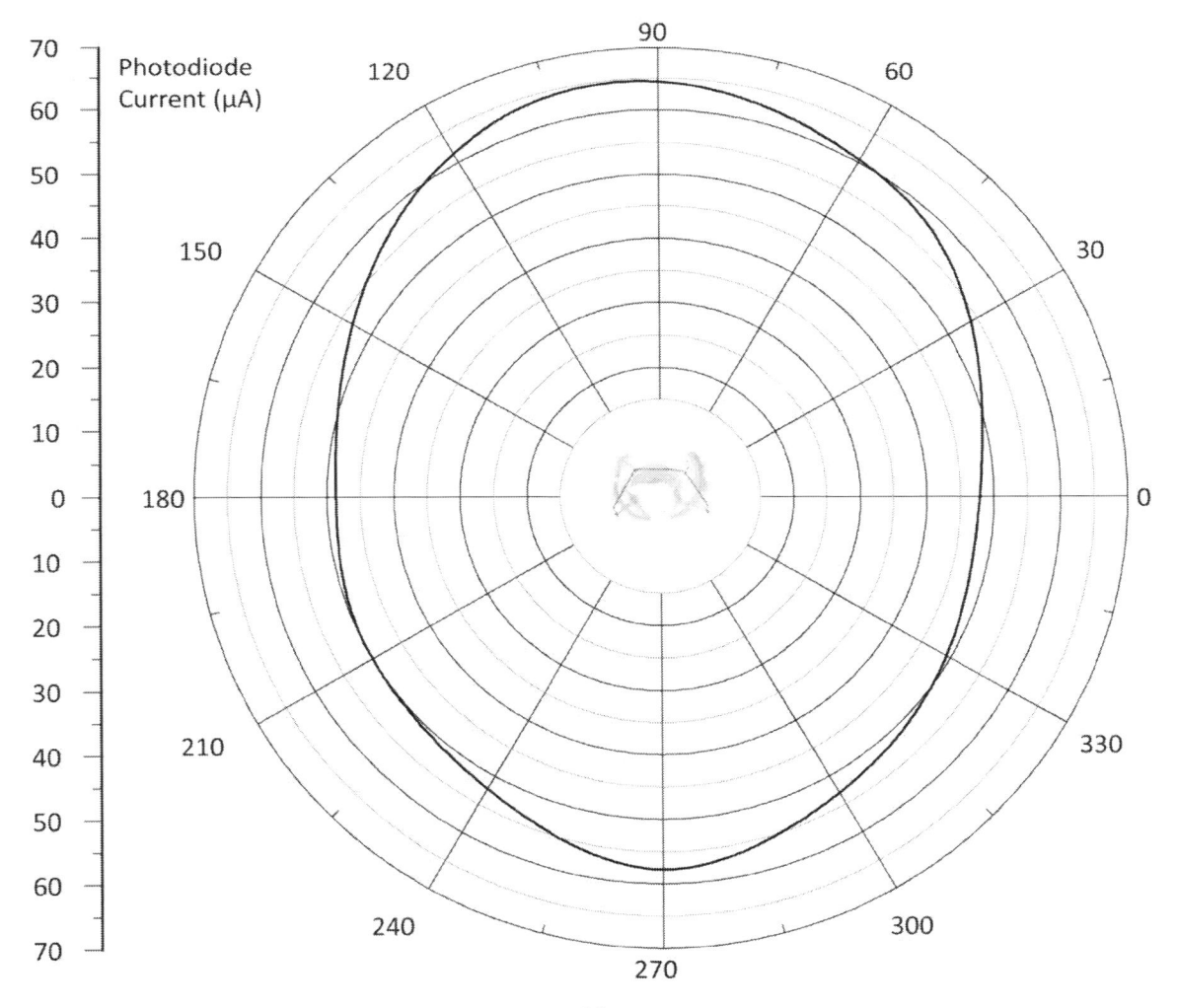

20 Lord Kelvin's Bridge

Measuring the resistance of a resistor is simple. You isolate it from the rest of the circuit and attach it to a digital multimeter. As an exam question, I could rephrase it as, *'how do you measure the resistance of a resistor?'* The answer, *'attach it to a digital multimeter'*, would be awarded 'nil points'. It's not that you need a humour bypass to be a physicist; it's just that physics is more than simply getting a result. You must know how to do it from first principles; there must be people who understand how to design a digital multimeter to measure resistance. This project investigates a simple, yet clever way of measuring resistance.

The **Wheatstone Bridge** is familiar to you from Higher physics. The plan is to start with the Wheatstone Bridge, follow it with a variation called the Carey-Foster Bridge, and finish with the Kelvin Bridge. You can use dc supplies throughout, though if you have more time for further experiments you can easily extend the project to cover ac supplies and measure capacitance and/or inductance.

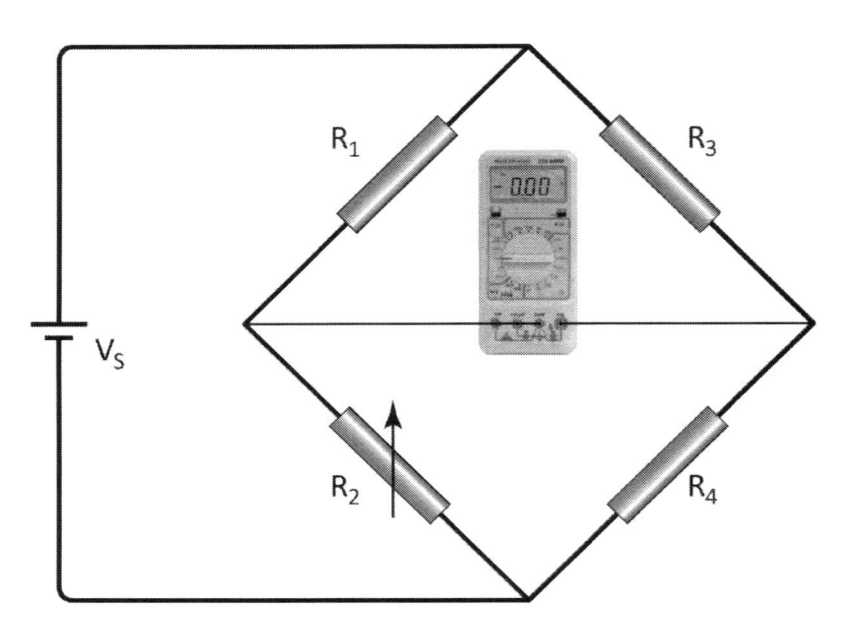

Set up the Wheatstone Bridge as in the diagram above and use it to measure the resistance R_1 of several resistors: large ($\approx$ 50kΩ), medium ($\approx$1kΩ) and small ($\approx$1Ω) resistances. The pd across the voltmeter (a digital multimeter humbled as a null indicator) is given by:

$$V = \left(\frac{R_4}{R_3 + R_4} - \frac{R_2}{R_1 + R_2} \right) V_S \quad \Rightarrow \quad V = 0 \text{ when } \frac{R_4}{R_3 + R_4} = \frac{R_2}{R_1 + R_2}$$

Cross multiply and check that this is equivalent to the usual balance condition. To take it beyond a mere repeat of Higher work, you should do a decent uncertainty analysis. You have two options: the easy way and the picky way. The easy way uses the balance condition to derive the uncertainty:

$$\frac{R_1}{R_2} = \frac{R_3}{R_4} \quad \Rightarrow \quad R_1 = \frac{R_2 R_3}{R_4} \quad \Rightarrow \quad \left[\frac{\Delta R_1}{R_1} \right]^2 = \left[\frac{\Delta R_2}{R_2} \right]^2 + \left[\frac{\Delta R_3}{R_3} \right]^2 + \left[\frac{\Delta R_4}{R_4} \right]^2$$

The contributions to the fractional uncertainties of each resistance are of equal importance and can be taken from the colour banding of each one (with permission, you might need to unscrew a cover to gain access). The problem with the easy way is that it assumes no uncertainty in the null reading of the voltmeter. It will have a reading uncertainty of ±½ the last digit. There might also be a systematic uncertainty (very uncommon in digital meters, but an analogue meter might not be accurately zeroed). The first equation in this topic shows that the unknown resistance R_1 depends upon five variables (R_2, R_3, R_4, V and V_S), so the expression for the uncertainty in R_1 should be made up of five terms. For students who refuse to be intimidated by a bit of maths, it's:

$$(\Delta R_1)^2 = \left[\frac{R_3}{R_4}\right]^2 (\Delta R_2)^2 + \left[\frac{R_2}{R_4}\right]^2 (\Delta R_3)^2 + \left[\frac{R_2 R_3}{R_4^2}\right]^2 (\Delta R_4)^2 + \left[\frac{R_2}{V_S}\left(1+\frac{R_3}{R_4}\right)\right]^2 (\Delta V)^2 + \left[\frac{VR_2}{V_S^2}\left(1+\frac{R_3}{R_4}\right)\right]^2 (\Delta V_S)^2$$

The second experiment is a modification to the Wheatstone Bridge, called the **Carey-Foster Bridge**. This was devised by the Professor of Natural Philosophy at Strathclyde University (known as Anderson's University in the early 1860's). If your school has a bought-in meter bridge, it will look like the one below. Wooden base, wide

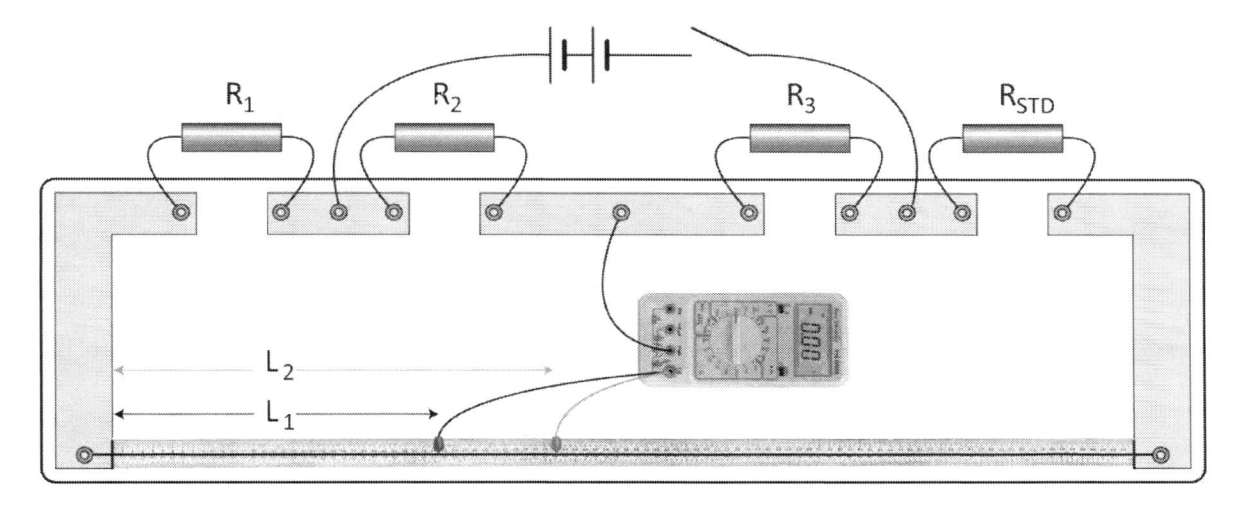

copper strips instead of wires, connection posts, and four gaps. Think of the resistance of the length of wire L_1 as an addition to R_1, and the remaining length as an addition to resistance R_{std} (when you're practising, this should be a standard resistance box – the type with decade selectors or brass plugs). This arrangement is the Carey-Foster Bridge. But it's more than just another way of doing the same thing as a Wheatstone Bridge; it's designed to measure the *difference* in resistance of two similar resistors, R_1 and R_{std}.

Its operation is simple. With the resistors as positioned in the diagram, the balance point L_1 is recorded on the meter stick. The resistors R_1 and R_{std} *are then swapped*, and the new balance point at L_2 is recorded on the meter stick. Start with R_2 and R_3 both the same (say, 100Ω, but not too large). Select the 'unknown' resistor R_1 as one of the nominal values eg 68Ω (with a 5% or 10% tolerance), and set the standard resistance box at the nearest integer value. Determine the two balance points as outlined above, calculate R_1 as outlined below, then practise with different resistors.

Your Report should derive the expression for the difference in resistance in terms of these two balance points. Here's how it's done. The metre length of wire has resistance σ ohms (pronounced 'sigma'). This is the same as saying that the resistance of the wire per unit length is σ Ωm^{-1}. So the length L_1 will have resistance σL_1 ohms (a 0.4m metre length of σ = 5 Ωm^{-1} wire will have resistance 5 x 0.4 = 2Ω). Start with the two balance equations:

$$\frac{R_2}{R_3} = \frac{R_1 + \sigma L_1}{R_{std} + \sigma(1-L_1)} \qquad \frac{R_2}{R_3} = \frac{R_{std} + \sigma L_2}{R_1 + \sigma(1-L_2)}$$

At this point, everyone (even Field's Medallists) would equate the right hand sides. You can get the answer this way, but the algebra comes thick and heavy. Instead, we use the sort of trick that makes you feel inadequate; add one to both sides.

$$\frac{R_2}{R_3}+1=\frac{R_1+\sigma L_1}{R_{std}+\sigma(1-L_1)}+1 \qquad \frac{R_2}{R_3}+1=\frac{R_{std}+\sigma L_2}{R_1+\sigma(1-L_2)}+1$$

Form common denominators:

$$\frac{R_2+R_3}{R_3}=\frac{R_1+\sigma L_1+R_{std}+\sigma(1-L_1)}{R_{std}+\sigma(1-L_1)} \qquad \frac{R_2+R_3}{R_3}=\frac{R_{std}+\sigma L_2+R_1+\sigma(1-L_2)}{R_1+\sigma(1-L_2)}$$

Tidy up the numerators:

$$\frac{R_2+R_3}{R_3}=\frac{R_1+R_{std}+\sigma}{R_{std}+\sigma(1-L_1)} \qquad \frac{R_2+R_3}{R_3}=\frac{R_{std}+R_1+\sigma}{R_1+\sigma(1-L_2)}$$

The denominators must be equal: $\qquad R_{std}+\sigma(1-L_1)=R_1+\sigma(1-L_2)$

Finally, tidy up to get the result: $\qquad R_1-R_{std}=\sigma(L_2-L_1)$

In words, the difference in resistance is proportional to the distance between the balance points. We just need to know the resistance per unit length of the wire. You can do this by identifying the metal, measuring its gauge and looking up tables. Or you can carry out a simple experiment using standard resistance boxes for both R_1 and R_{std}. Try a difference of 1Ω and measure the two balance points. Repeat for other combinations with a difference of 1Ω. For uncertainties, use absolute values:

$$\left(\Delta(R_1-R_{std})\right)^2=\left(\sigma\Delta L_1\right)^2+\left(\sigma\Delta L_2\right)^2$$

Once you've got it calibrated, measure the resistances of a bag of resistors of the same nominal value (eg. 39Ω) and plot a histogram of the distribution of values. If you've still more time, use it as a resistance thermometer!

We finally reach **Lord Kelvin's Double Bridge** with the third experiment. While Carey-Foster was Professor at Strathclyde University, Kelvin (William Thomson) was Professor at Glasgow University. It seems that the bridges they built in the east and west ends of the city have been forgotten by today's inhabitants! Kelvin's Bridge is an elegant and accurate way of comparing two small resistances. I say 'comparing' because, like all bridges, it doesn't actually measure the resistance; you only get a comparison with a known resistance (so the bit at the start of this section about measuring resistances with a Wheatstone Bridge is false). This is why at least one of the resistors is a standard resistance box. I also said 'small'. The Wheatstone Bridge can be used to 'measure' resistances of all sizes, but problems arise with small resistances. The resistance of the connections and of the wires cannot be neglected. This is one of the nice features of the Carey-Foster Bridge; if you include these other small voltage drops in the calculations at the top of this page, you discover that they cancel out by the final stage. Kelvin's version of the bridge, though, is especially suited to very small resistances. It can measure resistances of less than 0.0001Ω (you can buy one claiming down to 10nΩ!!!)

Kelvin's Bridge can be drawn in different ways, but they are all electrically the same. Here is the traditional one:

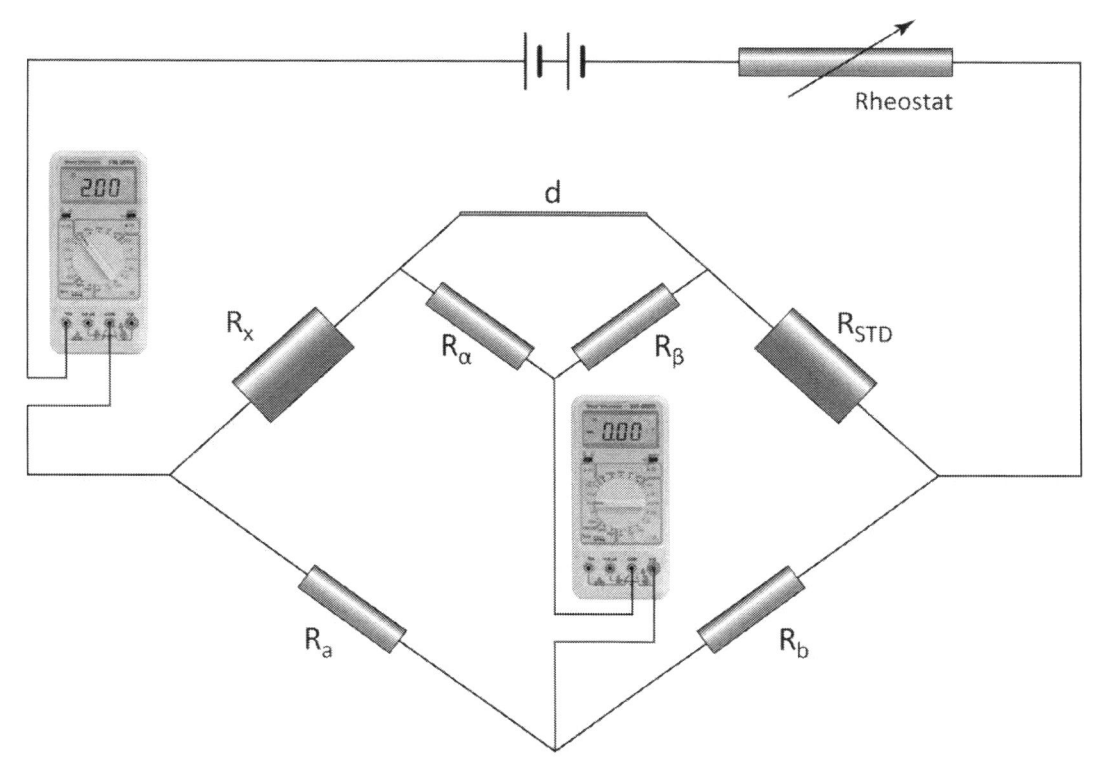

Measuring small resistances can lead to large currents (there is a low resistance path through the top arm), so a suitably thick rheostat is placed in series with the supply to limit it, with an ammeter in series as monitor. The first thing you notice is the extra pair of resistors R_α and R_β, hence the name 'double bridge'. The part labelled 'd' is a connection made with as low a resistance as possible (it's called the 'yoke'). Although the diagram is drawn with the wires spread out, in practise all the connections in the bridge section are made as short and secure as possible. Typical starting values would be: any combination of 10Ω, 100Ω, or 1000Ω for R_a, R_b, R_α and R_β (subject to the ratios constraint below), and about 10Ω for R_{std}. The resistors R_α and R_β shouldn't be too big or you lose the sensitivity of the voltmeter. At balance (voltmeter reading zero):

$$\frac{R_X}{R_{STD}} = \frac{R_a}{R_b} + \frac{R_d}{R_{STD}} \left(\frac{R_\beta}{R_\alpha + R_\beta + R_d} \right) \left(\frac{R_a}{R_b} - \frac{R_\alpha}{R_\beta} \right)$$

If the right hand bracket is zero, this will give the same relation as the Wheatstone Bridge. This corresponds to the ratios R_a/R_b and R_α/R_β being equal. Select resistance values accordingly.

After setting up for the first time, measure resistances in single figures using a ratio of 1:10 with a standard resistance box in 1's up to 100Ω. As you get more confident, try to measure smaller resistances (as a challenge, how about measuring the resistivity of magnesium?) The standard resistor R_{STD} needs to have a much smaller resistance this time, yet still be variable and capable of carrying a decent current. A 1m length of 24SWG copper wire has a resistance of about 0.06Ω and will carry a current of several amps without additional cooling. Try using this with a metre stick and sliding contact. You'll discover lots of good experimental physics with this project; just don't give up (and keep an eye on these currents)!

Coming up with your own Project Idea

This is the rock climbing equivalent of a first ascent. You could be in uncharted territory; never sure what's around the next corner, and liable to fall off at any moment. The secret is to climb easy ground. Don't be unrealistic. You may have an interest in, say, mountain biking, where you can bring particular knowledge of the subject and provide the equipment yourself. If you've got sensible, workable ideas, get the plan down on paper and go for it. But not all ideas are sensible. Here are a few areas to be wary of.

Anything **nuclear** (a popular choice) will have little hope of coming to fruition, partly because of Health & Safety regulations, and partly because there's little you can do (universities have things like solid state energy spectrometers but the project would be too much of the 'pushing buttons' type rather than 'hands-on').

Another common choice to be wary of is the overly '**electronic**' project. A Report on the properties of a transistor or MOSFET will tip the SQA marker over the edge, and any projects involving complicated circuits are liable to move outside physics into technological studies. There are a number of teachers at loose in Scotland who are electronics enthusiasts and their enthusiasm can be infectious. Be on guard!

Then there's **Special Relativity**. Forget it, unless you want to repeat the Michelson Morley experiment with two measurements about 6 months apart entailing an early start and late finish. Measuring the speed of light using rotating toothed wheels and lasers isn't testing Special Relativity. Using basic equipment, it is a challenging and interesting project in its own right. An even worse choice would be anything to do with General Relativity. If your life has been characterised by unbroken success and you feel like a change, then look no further.

Most **astronomy** experiments either consist of cutting out bits of cardboard or having access to a decent telescope; there is a lack of experiments at the level of AH. You're up against the problem of collecting enough numerical data to analyse. There may be possibilities for an optics based investigation with telescopes (see the bottom of page 37 for an example with focal lengths), though this lacks some of the excitement of the original topic. But this is an interesting area and unusual ideas like measuring light pollution might be possible. A modern digital SLR camera attached to a modest telescope (or just a zoom lens) combined with free software like DeepSkyStacker, opens up new possibilities.
Try this website, http://cse.ssl.berkeley.edu/AtHomeAstronomy/

Universities are keen to forge links with local schools and a visit could be arranged through your teacher. Tough luck if you live in Skye perhaps, but that's just Scotland's lopsidedness. Good facilities are often available at universities, but it also means that unless you can borrow equipment, you'll need to make repeat visits to do the experiments. Remember that you are representing your school, so try to leave a good impression.

Quite often, choosing your own project comes down to an equipment problem, but if you can beat that one, you could be on to a winner. So much of SQA marker's time is taken up assessing the same old favourites, that novelty projects have a head start. But, if your grammar is poor, the physics is lousy and the report looks like it wrapped up a fish supper, then novelty won't help you. Just make sure you come across as intelligent.

The Daybook 3

This is the neglected part of the exercise. And I think it's due to our inability to sell it to the students. How do you answer a student who asks, 'what's the point of having all the same stuff in the Daybook when it's already in the Investigation Report?' The easy answer is that they have to do it because it's an SQA requirement. The student who likes a degree of independence will only wear that jacket with reluctance. The alternative answer is to make the case that it's good practice to take notes and record observations **at the time of the experiment**. Our student with the independent mind will be more ready to accept this one, but then ask why it should be examinable. However, **it will be assessed by the teacher**, so what should go into the Daybook?

A jotter of the thicker variety will do (it doesn't have to be as big as A4, though that will do nicely). Loose leaf folders aren't as good as jotters since the ability to remove pages without leaving any trace reduces the sense of excitement and *reportage* of the firsthand account. If you are the more zealous type of student, then don't waste your efforts on the daybook; keep that for the Investigation Report. It's what's **in** the Daybook that's important, not how it's presented.

Your teacher will look for two main items; the planning of the Investigation, and the recording and analysis of the results (these are called Outcome 1 and Outcome 2).

Outcome 1

....a record is maintained in a regular manner....   Turning out rubbish on the hour every hour is 'regular'. Basically, it means write something in it every experimental session (put in the date), and keep it legible and readable for your teacher. Don't make it like Galileo's notebooks with scribbles at all angles all over the place (see this web URL for examples, http://www.mpiwg-berlin.mpg.de/Galileo_Prototype/MAIN.HTM).

....techniques and apparatus are appropriate...   Experimental physics is about measuring something. You will have to come up with a plan of attack to do this; which method, and what equipment to use. For example, as part of an Investigation you might have to measure the distance between the goal posts at the opposite ends of a football field. How would you do it? How accurate do you have to be? Let's say as accurate as possible. Would a metrestick be appropriate? How about a cloth tape-measure from the PE department? How about a bit of geometry using a baseline and angles? Put down all these ideas in your daybook and discuss the merits of each one, finally choosing the one you think is most appropriate. Then there's **using** the equipment. During Standard Grade in 4th year you might remember dropping plasticene from various heights in the lab. Some pupils measured the starting height using one metrestick and a horizontal finger, others used the same metrestick in a more appropriate manner. Assessment of this outcome is either pass or fail; to fail it you'd have to be the 'Inspector Clouseau' of experimental physics.

Outcome 2

....collection of data is carried out with accuracy...

They don't really mean 'accuracy' here (since that depends on the equipment you've been able to get), they mean with due care and attention. In other words, you've done the best you can with the equipment available.

....measurements are recorded in an appropriate format...

Tables with proper headings and units. If you record lots of data through a computer and output tables from the likes of Excel, then it's okay to paste it into your daybook (just make sure that units and headings are included). Use the recommended SI units for headings, eg seconds rather than minutes. If you finish a series of readings and decide that you need more readings within a certain range, just extend the table downwards. The table in the Report can have the readings in the proper order.

....information is analysed and presented in an appropriate format...

If an equation is being used, give one example showing substitution of the relevant numbers. Any graphs don't have to be as neat and tidy as in the Report, but they should have the proper scale, titles and units, and they should be plotted accurately (if you do it at this stage, it saves time later on with error bars). A graph is a thing which is plotted after the experiment, so drawing a really nice accurate graph and photocopying it for both the Daybook and Report is okay.

....uncertainties are treated appropriately...

Uncertainty calculations can be time consuming and tedious. Use the Daybook as your rough copy and transfer it neatly for the Report. Here is a row from a table showing a typical example:

Voltage (volts)	Current (amperes)	Resistance (ohms)
6.8 ± 0.1	0.27 ± 0.02	25.2 ± 1.9

Remember to include a note of any calibration uncertainties in the equipment you're using (and where you got the information from).

The Daybook isn't something to get uptight about. Make it useful by collecting book references and website URL's, and put in notes to yourself like ....**must include this in the Report**....... Write down your ideas during the planning stage, record your results every time you do an experiment, and use it for uncertainty calculations. That will be sufficient for the teacher to pass it. Keep your big effort for the Report.

The next few pages show samples from the Daybook of a student who received a final mark of 20/25.

Underlying Physics

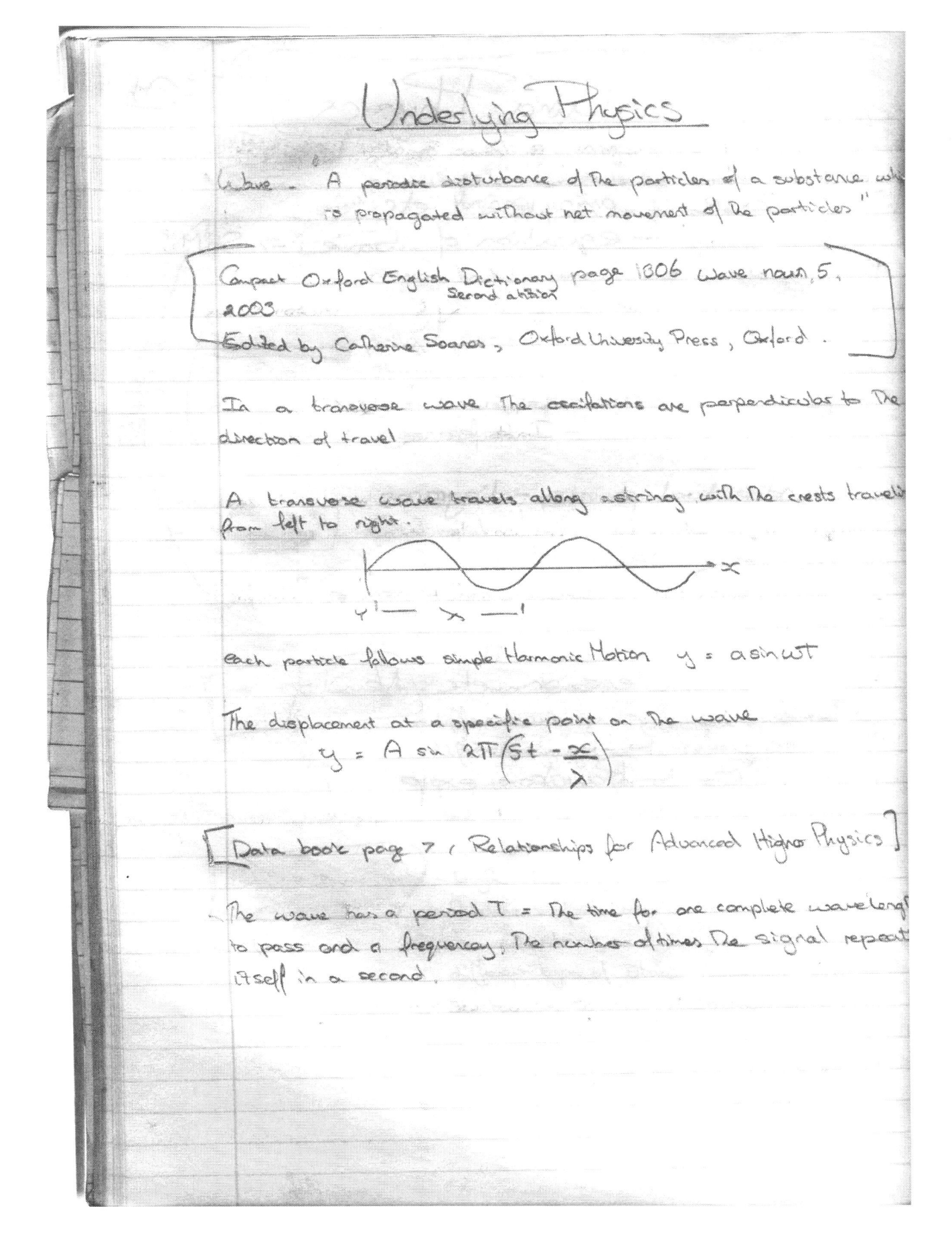

"Wave – A periodic disturbance of the particles of a substance which is propagated without net movement of the particles"

[Compact Oxford English Dictionary page 1306 wave noun, 5, Second edition, 2003
Edited by Catherine Soanes, Oxford University Press, Oxford.]

In a transverse wave the oscillations are perpendicular to the direction of travel.

A transverse wave travels along a string with the crests travelling from left to right.

Each particle follows Simple Harmonic Motion $y = a \sin \omega t$

The displacement at a specific point on the wave
$$y = A \sin 2\pi \left(ft - \frac{x}{\lambda} \right)$$

[Data book page 7, Relationships for Advanced Higher Physics]

The wave has a period T = the time for one complete wavelength to pass and a frequency, the number of times the signal repeats itself in a second.

17/1/07 Investigation Title: String Vibrations

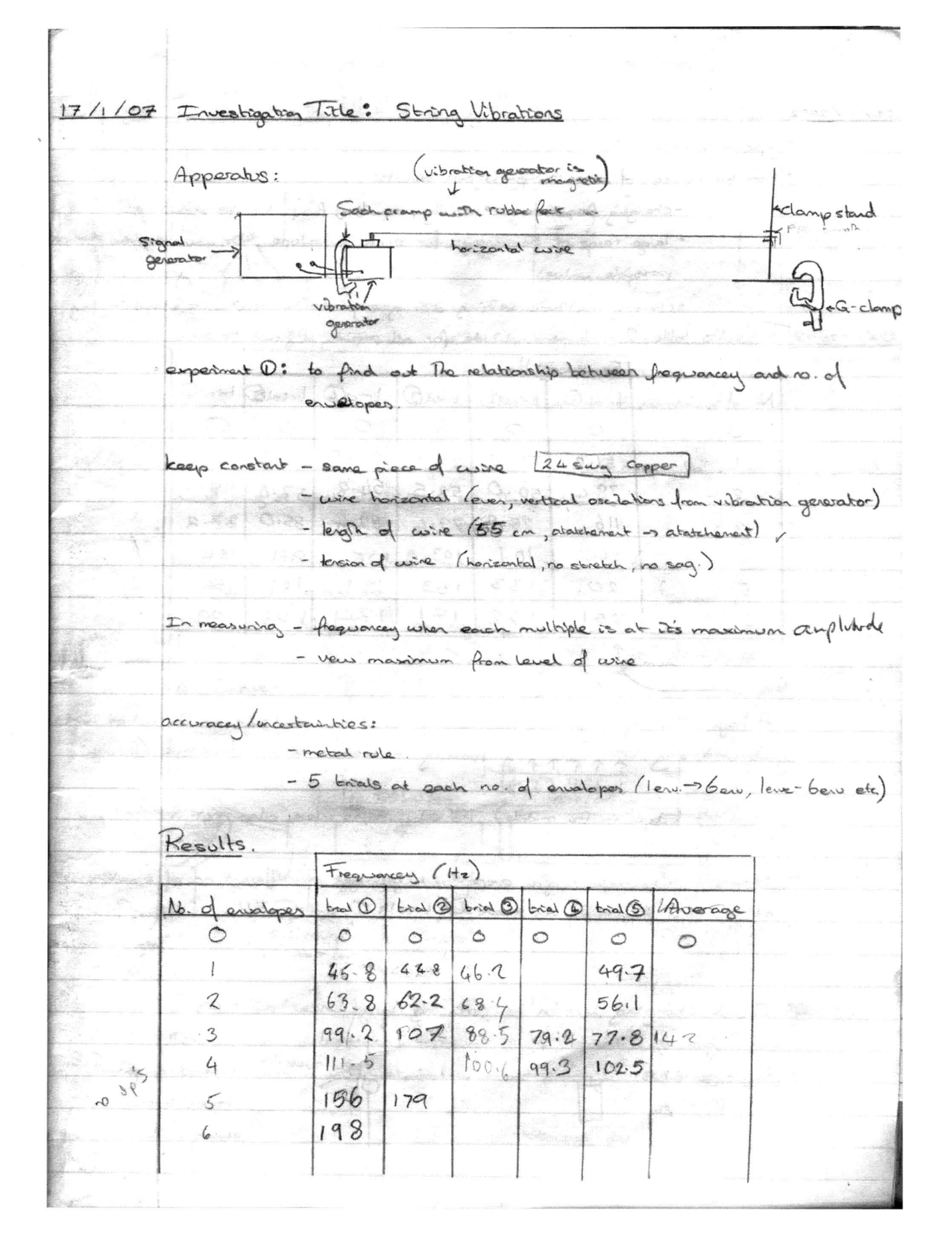

Apparatus:

(vibration generator is magnetic)

Sash preamp with rubber feet — horizontal wire — clamp stand
Signal generator → vibration generator, G-clamp

experiment ①: to find out the relationship between frequency and no. of envelopes.

keep constant — same piece of wire [24 swg copper]
- wire horizontal (even, vertical oscillations from vibration generator)
- length of wire (55 cm, atatchment → atatchment) ✓
- tension of wire (horizontal, no stretch, no sag.)

In measuring — frequency when each multiple is at its maximum amplitude
- view maximum from level of wire

accuracy/uncertainties:
- metal rule.
- 5 trials at each no. of envelopes (1env. → 6env, 1env - 6env etc)

Results.

No. of envelopes	Frequency (Hz)					
	trial ①	trial ②	trial ③	trial ④	trial ⑤	Average
0	0	0	0	0	0	0
1	45.8	44.8	46.2		49.7	
2	63.8	62.2	68.4		56.1	
3	99.2	107	88.5	79.8	77.8	142
4	111.5		100.6	99.3	102.5	
5	156	179				
6	198					

Uncertainties

Experiment 1

data table 1

% uncertainty	No. of envelopes	trial ①	trial ②	trial ③	trial ④	trial ⑤	Average
	0	0	0	0	0	0	0
(13.8)	1	51.2	26.0	45.8	29.1	29.3	36.3 ± 5.0
9.4	2	78.4	50.0	63.8	54.8	57.0	60.8 ± 5.7
8.7	3	116	75.8	99.2	82.4	85.0	91.7 ± 8.0
10.4	4	160	101	111	105	99	115 ± 12
9.3	5	201	133	156	131	129	150 ± 14
(7.7)	6	250	172	198	174	178	194 ± 15

Frequency for envelope at its maximum (Hz)

* Scale reading uncertainty in meter (sig gen) = ± 0.1 Hz (up to 100 Hz)
  ± 1 H (over 100 Hz)

* Calibration uncertainty in sig gen ⇒ calibrated to — insignificant.

Ignore scale reading uncertainty — less than 1/10 of random uncertainty.

* The random uncertainty was calculated using random uncertainty = $\frac{max - min}{n}$

- Perhaps changing tension of string is changing throughout experiment causing these large random uncertainties (between 7.7 and 13.8%).

- Perhaps frequency was increased too fast : waves had no time to stabilise

①

Frequency (Hz)	Wave at max (mm)	Amplitude (mm)
46.5	5	2.5
46.0	5	2.5
45.5	4	2
45.0	4	2
44.5	5	2.5
44.0	5	2.5
43.5	5	2.5
43.0	5	2.5
42.5	4	2
42.0	4	2
41.5	4	2
41.0	4	2
40.5	4	2
40.0	4	2
39.5	3	1.5
39.0	3	1.5
38.5	3	1.5
38.0	2	1
37.5	2	1
37.0	2	1
36.5	2	1
36.0	2	1
35.5	0 (string vibrating along whole length)	0
35.0	0 "	0
34.5	1	0.5
34.0	1	0.5
33.5	0 "	0
33.0	0 "	0
32.5	1	0.5
32.0	1	0.5
31.5	2	1

②

Frequency (Hz)	Wave at max (mm)	Amplitude (mm)
31.0	9 (2)	4.5
30.5	8	4
30.0	7	3.5
29.5	7	3.5
29.0	6	3
28.5	6	3
28.0	6	3
27.5	5	2.5
27.0	5	2.5
26.5	4	2
26.0	4	2
25.5	3	1.5
25.0	3	1.5
24.5	2	1
24.0	2	1
23.5	1	0.5
23.0	1	0.5
22.5	0 (string vibrating whole length)	0
22.0	0 "	0
21.5	0 "	0
21.0	1	0.5
20.5	1 "	0.5
20.0	0 "	0
19.5	0 "	0
19.0	0 "	0
18.5	0 "	0
18.0	1	0.5
17.5	0 "	0
17.0	1	0.5
16.5	0 "	0
16.0	2	1
15.5	4	2
15.0	14	7

1 envelope

Method:

- Slowly increase frequency from 0Hz — keeping input rms voltage constant at 1.5V
- Using the steel rule to measure amplitude at its maximum record the frequency for each no. of envelopes.
- Repeat, starting again from 0Hz.

Results:

No. of envelopes	Frequency in Hz					Average	Scale Read Uncertainty
	trial①	trial②	trial③	trial④	trial⑤		
0							
1	26.3	27.2	27.4	27.5	27.0	27.1	±0.1
2	54.8	54.4	55.2	55.2	55.1	54.9	±0.1
3	84.1	85.4	82.9	84.5	83.3	84.0	±0.1
4	105.9	106.6	103.9	108.0	108.2	106.5	±0.1
5	128	127	123	128	126	126	±1.0
6	155	154	151	157	155	154	±1.0

Random uncertainty = $\frac{max - min}{n}$

① = $\frac{27.5 - 26.0}{5}$ = ±0.3   ④ = $\frac{108.0 - 103.9}{5}$ = ±0.8

② = $\frac{55.2 - 54.2}{5}$ = ±0.2   ⑤ $\frac{128 - 123}{5}$ = ±1.0

③ = $\frac{85.4 - 82.9}{5}$ = ±0.5   ⑥ $\frac{157 - 151}{5}$ = ±1.2

Conclusion: As graph gives a straight line through origin (within uncertainty) frequency is directly proportional to no. of envelopes.

$$F \propto n$$

Evaluation

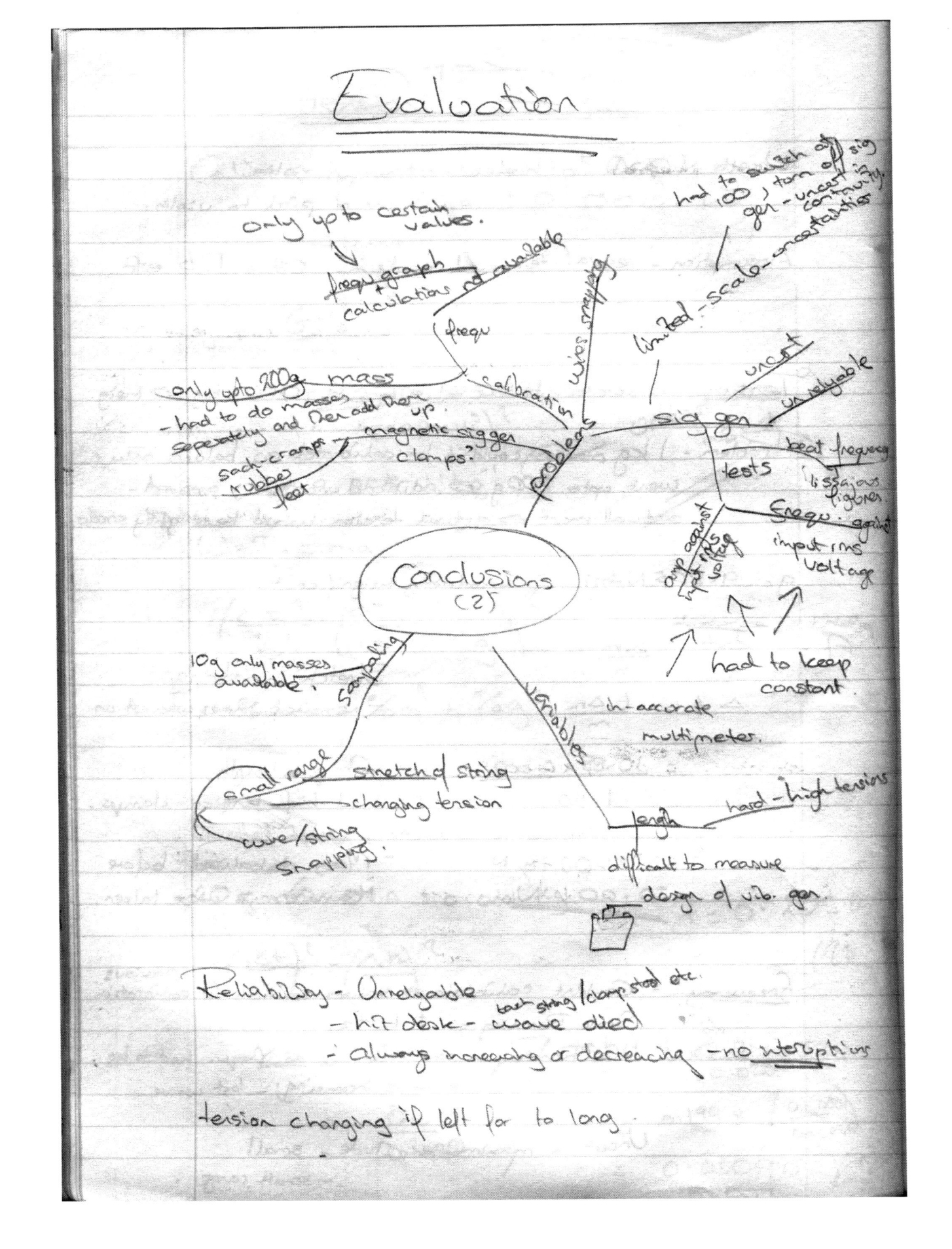

Mind map centered on **Conclusions (2)**:

- **problems** → frequ → graph + calculations not available (only up to certain values)
  - mass → only up to 200g — had to do masses separately and then add them up
  - clamps? → sash-cramps, rubber feet, magnetic rigger clamps
  - calibration → wires snapping
- **sig gen** → had to switch off sig gen — turn off continuity, uncertainty
  - limited — scale — uncertainties
  - uncert
  - unreliable
- **tests** → beat frequency, Lissajous figures, frequ against input rms voltage, amp against input rms voltage
  - had to keep constant
- **sampling** → 10g only masses available
- **small range** → stretch of string, changing tension, wave/string snapping
- **variables** → in-accurate multimeter
  - length → hard — high tension, difficult to measure
  - design of vib. gen

Reliability – Unreliable — broke string/clamp stand etc.
- hit desk — wave died
- always increasing or decreasing — no interruptions

tension changing if left for too long.

Writing the Report

You will never meet the SQA external Assessor who will mark your report. You will not be able to explain to the Assessor how good you are at experimental physics and that writing reports isn't your 'thing'. The Assessor will mark you out of 30 based solely on the Report in front of him.

Some teachers (including me), think this is educationally unsound, and that the Assessor should visit the school to interview the candidate (it used to be like that in the last century). Nobody is disputing that report-writing is an important skill, but should we equate an accomplished experimenter who writes a mediocre report, with a mediocre experimenter who writes a dazzling report? After all, it is supposed to be a Physics Report, rather than a marketing document.

Returning to the 'real' world, the total mark of 30 for the Report is split into 6 main sections. Being prescriptive like this removes some of the element of judgement which an experienced marker builds up over the years, but also prevents an inexperienced marker from making gross errors in assessing the Report. If an experienced Assessor holds a scruffy looking report in his hands, but detects the signs of a good experimenter, they can only use that judgement to a very limited extent. For example, don't laugh, credit for expressing book references in the correct format. Miss them out and you don't get the marks; no matter how good you are at Physics or how experienced is the marker.

So here's my advice for writing the Report:

......use the Marks Allocation sheet as a tick list......

The next page shows you the details of the marks allocation. Keep it beside you at all times when writing the Report. When you've finished the first draft, check that you've addressed every one of these points.

How many Words, How many Pages?

The SQA suggests between 2500 and 4500 words, hinting that reports 'using considerably fewer words than the maximum', can also gain full marks. On the number of pages, 20 would be a bit light and 80 would be too heavy. Present it single-side.

Can it be Handwritten?

Yes. Most students will use computers to write the Report, typically Microsoft Word and Excel for the text, tables and graphs. Graphs can often be unsatisfactory using computer software; draw by hand if necessary.

Marks Allocation	Marks
• Abstract	1
• Introduction	4
• Procedures	7
• Results	8
• Discussion	8
• Presentation	2
Total marks	30

Use of Language

There is a tradition which demands that all scientific and technical writing be expressed without using personal pronouns (like, I, we, you, she). You shouldn't write things like, *'I did the experiment….'*. It should be, *'An experiment was performed…..'*. The first way sees the world from the inside looking out and the second way sees it from the outside looking in. A scientific observer is supposed to be detached and impartial and so should remove himself from the Report. **The SQA discourages the use of pronouns in a Report**. Or as they would put it, *use the passive voice, not the personal voice*. Here are three examples of scientific language from top physicists. The SQA would disapprove of the middle one.

1. **arXiv: gr-qc0712.1649v2   Richard K Obousy and Gerald Cleaver   2007**

A 1994 paper written by M. Alcubierre [2] demonstrated that, within the framework of general relativity, a modification of spacetime could be created that would allow a spacecraft to travel with arbitrarily large speeds. In a manner identical to the inflationary stage of the universe, the spacecraft would have a relative speed, defined as change of proper spatial distance over proper spatial time, faster than the speed of light. Since the original paper, numerous authors have investigated and built on the original work of Alcubierre. Warp drives provide a unique and inspiring opportunity to ask the question 'what constraints do the laws of physics place on the abilities of an arbitrarily advanced civilization' [21].

2. **On the Electrodynamics of Moving Bodies   A. Einstein June 30, 1905**

If we wish to describe the motion of a material point, we give the values of its co-ordinates as functions of the time. Now we must bear carefully in mind that a mathematical description of this kind has no physical meaning unless we are quite clear as to what we understand by ``time.'' We have to take into account that all our judgments in which time plays a part are always judgments of simultaneous events. If, for instance, I say, ``That train arrives here at 7 o'clock,'' I mean something like this: ``The pointing of the small hand of my watch to 7 and the arrival of the train are simultaneous events.''

3. **Reflections on the Fate of Spacetime   Edward Witten 1996**

Our basic ideas about physics went through several upheavals early this century. Quantum mechanics taught us that the classical notions of the position and velocity of a particle were only approximations of the truth. With general relativity, spacetime became a dynamical variable, curving in response to mass and energy. Contemporary developments in theoretical physics suggest that another revolution may be in progress, through which a new source of "fuzziness" may enter physics, and spacetime itself may be reinterpreted as an approximate, derived concept.

The Abstract

All scientific reports from universities and research institutes start with a very short, concise description of what it contains, called the 'abstract'. If your report contains a numerical result, it should be quoted in the abstract. In science the abstract is used to 'sell' the article to the reader, and is therefore considered to be very important. Scientists may also only read the abstract and the conclusion of a paper, so these two parts must convey clearly the central message of the report, including the main result and its implications. Here's an example of an abstract from the main physics hub at arXiv.org (with the title to let you know what it's about):

— — — — — — —

Measuring Star Formation in Local and Distant Galaxies

Abstract: *This article reviews measurements of star formation in nearby galaxies in the UV-to-Far IR wavelength range, and discuss their impact on SFR determinations in intermediate and high redshift galaxy populations. Existing and upcoming facilities will enable precise cross-calibrations among the various indicators, thus bringing them onto a common scale.*

— — — — — — —

Notice how sharp and concise is the language. Try to come up with one in the same style for your Report. The page will look a bit bare; that's okay. Here's an example from a Sound Investigation:

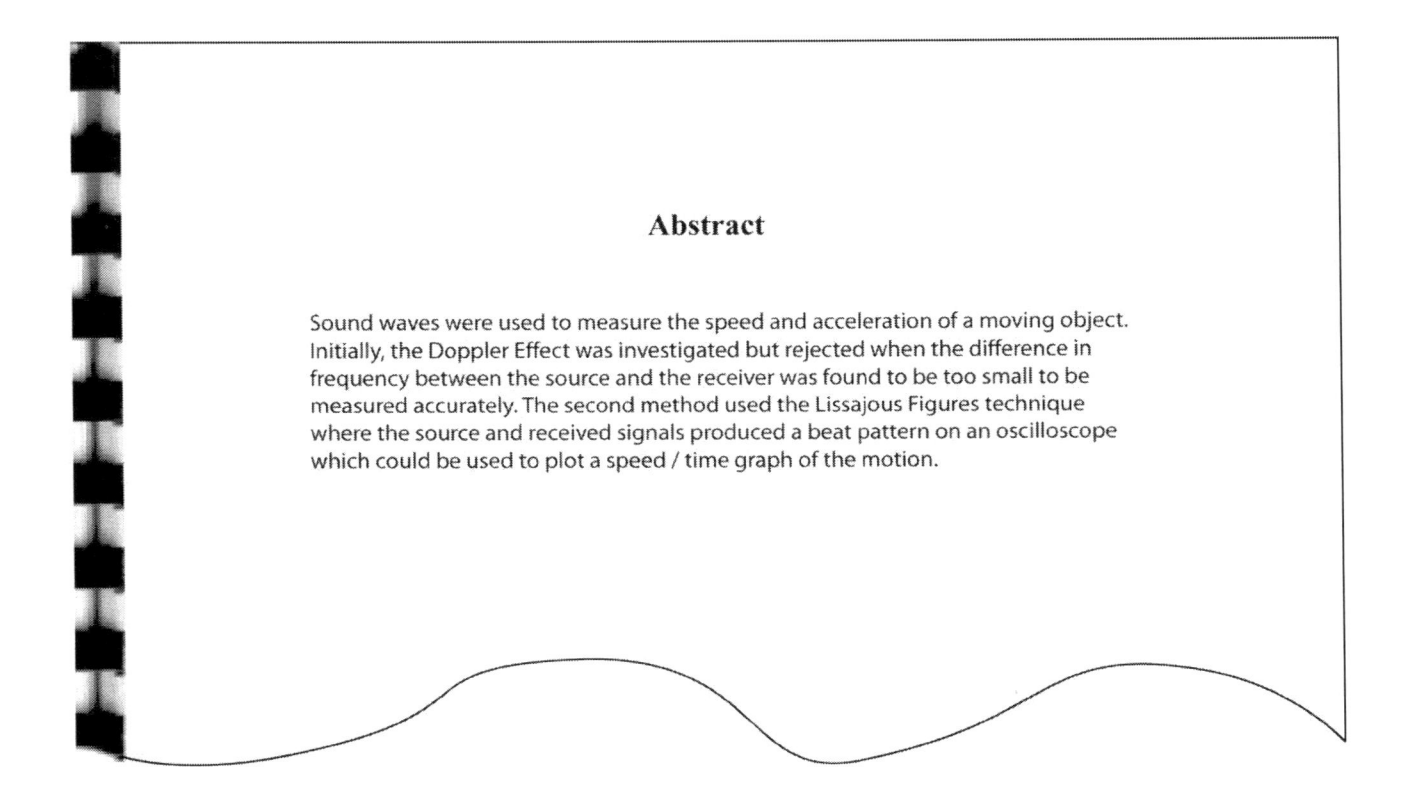

Abstract

Sound waves were used to measure the speed and acceleration of a moving object. Initially, the Doppler Effect was investigated but rejected when the difference in frequency between the source and the receiver was found to be too small to be measured accurately. The second method used the Lissajous Figures technique where the source and received signals produced a beat pattern on an oscilloscope which could be used to plot a speed / time graph of the motion.

Section 1   The Introduction

> - **Summary** stating the purpose and overall findings of the Investigation
> - Account of the **underlying Physics** that is:
>   — *relevant to the Investigation*
>   — *complete*
>   — *of a level appropriate to the demands of AH Physics*
>   — *more than just a simple repetition of coursework*

Summary

The summary should be a concise description of what the project is about, and include any results of the measurement of a physical quantity. To real scientists, placing it as the first item in the introduction will seem odd; the reason I suggest you do this is to pick up the available mark. Here is a good example of a summary:

Some of the properties of an ACME helium neon laser were investigated. The diffraction grating method was used to measure the wavelength of the laser light, giving a result of (625±10) nm. Laser beams also spread outwards slightly. Using simple geometry and a photodiode, the divergence of the beam emitted by the laser was found to be (0.06±0.01)°. An attempt was made to measure the power output of the beam by shining it onto the blackened surface of a small thermistor for a prolonged period of time. The result with its uncertainty lay outside the claimed power output of the laser given by the manufacturer, though of the same order of magnitude. Reasons are offered for the discrepancy and suggestions made for improving the experiment.

Try to write the summary with a 'cool' impersonal style. It forces you to make it sharp and snappy, rather than self-centred. Here's the same Investigation with a poor summary:

Well I measured a laser beam for three things. The wavelength which turned out to be 625±10nm and is quite a good answer. I then saw how much it spread out. This was 0.06°±0.01°. Finally I tried to measure its power but this was difficult and I didn't get a good answer.

This includes the same results as the good example, but makes no mention of the methods used. Remembering that you will never meet the external assessor, the good example suggests a student who has taken some care and thought over its preparation, while the poor example suggests a student whose Investigation is getting in the way of his social life.

Underlying Physics

This is the second subheading under 'Introduction'. Take a few pages over it, and let the physics unfold slowly. If you have an equation then show how it's derived, or if it's too complicated (like say in viscosity), give a proper reference to a book. Web references aren't quite so good since they depend upon the website being maintained, whereas a published book is permanent. Here is an example of part of a good write-up:

➤➤➤➤➤➤➤➤➤ EXAMPLE BEGINS HERE ⬅⬅⬅⬅⬅⬅⬅⬅

The Underlying Physics

A simple pendulum is defined as a swinging point mass, which is suspended from a string or rod of negligible mass. If a pendulum is set in motion so that it swings back and forth, then its motion will be periodic. The period of a pendulum is the time taken for it to complete one full cycle or oscillation. Another important quantity is the frequency of oscillation. The frequency f is the number of oscillations that occur per second and is related to the period by the formula:

$$f = \frac{1}{T}$$

This is a useful formula as it is the inverse of the period. Another quantity of periodic motion is the amplitude of oscillation. The amplitude is the point of maximum distance that the pendulum moves away from its equilibrium position. When a simple pendulum is set in motion it is displaced from this point of equilibrium. When this happens there is a restoring force created which pulls the pendulum back towards and past its equilibrium point. Once carried past the point of equilibrium the direction of the restoring force reverses so it is again acting towards the equilibrium position. The pendulum's motion has an acceleration that is always proportional to its displacement, but of opposite sign. This particular form of motion is known as Simple Harmonic Motion (SMH) and is important for understanding the underlying physics involved when investigating the period of a pendulum.

We begin by deriving the expression for the period of a simple pendulum of mass 'm' and length 'l':

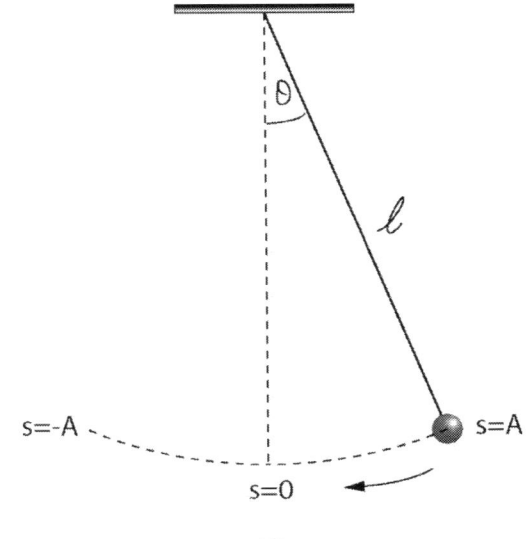

The pendulum of length '*l*' makes an angle 'θ' with the vertical and has amplitude 'A' There are two forces on the pendulum bob, the String Tension and the Weight of the bob.

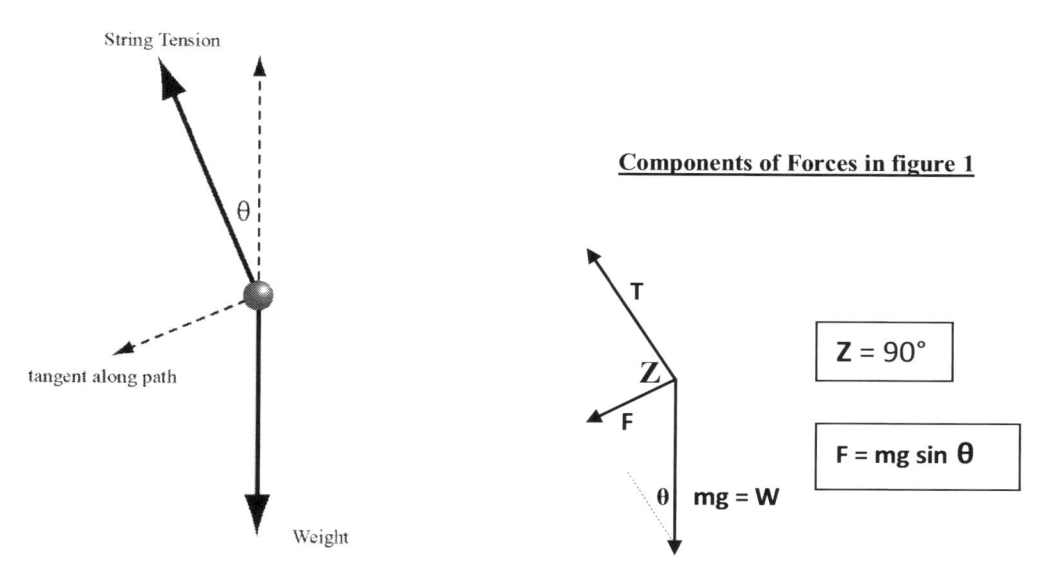

Figure 1: Diagram of Simple Pendulum and its associated Forces

Components of Forces in figure 1

$Z = 90°$

$F = mg \sin θ$

If the pendulum is displaced by an angle θ from the vertical, the bob experiences a restoring force due to gravity:

$$F = -mg \sin \theta$$

The negative sign is put in to show that the restoring force is in the opposite direction to the direction of increasing angle θ. Note that the motion of the bob is at right-angles to the string tension and hence does not receive a component from this force. We make the approximation that when the angle θ is small, **sin θ ≈ θ** (θ in radians).

The restoring force becomes:

$$F = -mg\theta$$

If the bob moves through an arc length '*s*', then:

$$s = l\theta$$

and we obtain:

$$F = -mg\frac{s}{l} \quad\quad\quad \text{formula (a)}$$

73

Since the restoring force is proportional to the displacement, the pendulum is obeying simple harmonic motion and therefore an example of the relationship **F** = − **ks**. Using Newton's 2nd Law, we obtain:

$$ma = -mg\frac{s}{l}$$

The masses cancel (hence its motion is independent of mass) and we use the derivative expression for the acceleration:

$$\frac{d^2s}{dt^2} = -\frac{g}{l}s$$

The solution is a sine or cosine. We use the sine ($s = 0$ at $t = 0$):

$$s = A\sin\omega t$$

Substituting:

$$\frac{d^2(A\sin\omega t)}{dt^2} = -\frac{g}{l}A\sin\omega t$$

$$-\omega^2 \sin\omega t = -\frac{g}{l}\sin\omega t$$

$$\omega^2 = \frac{g}{l} \quad \Rightarrow \omega = \sqrt{\frac{g}{l}}$$

$$\Rightarrow \boxed{T = 2\pi\sqrt{\frac{l}{g}}} \qquad \text{formula (1)}$$

EXAMPLE ENDS HERE

This was the first part of her description of the underlying physics, and as such is just a repetition of the coursework in SHM. It was used as a lead-in to the main section of theory, where she went on to derive the expression for the period of a physical pendulum (something which is less familiar) using the moment of inertia. No references were needed at this stage since the expression for the period was derived from first principles. A few pages further on, she quoted the equation for the period of a simple pendulum where the angle of oscillation is large and the approximation $\sin\theta \approx \theta$ cannot be made. This required a book reference.

Section 2   Procedures

- *Labelled **diagrams** and/or descriptions of apparatus, as appropriate.*

- *Clear **descriptions** of how the apparatus was used to obtain experimental readings.*

- *Appropriate level of demand; factors to be considered include:*
    - *range of procedures*
    - *repeatability*
    - *control of variables*
    - *accuracy*
    - *originality of approach and/or experimental techniques*
    - *degree of sophistication of experimental design and/or equipment*

This is the most 'wordy' part of the Report; decent language skills are useful. Two different people could have the same information about an experiment in their brains, but one might write a flowing, relevant description in a nice logical sequence, whereas the other writes it like a badly composed text message. But no matter how bad you are at composing an essay (and it's a bit like that, but without the vanity), you can produce well drawn, labelled diagrams. Manage that and you've done the first bit; two marks in the bag. Here's an example of a poor diagram which is all too common. The student has used a very basic computer drawing program (probably Word). They should have drawn it by hand, or taken a digital photograph and added labels later.

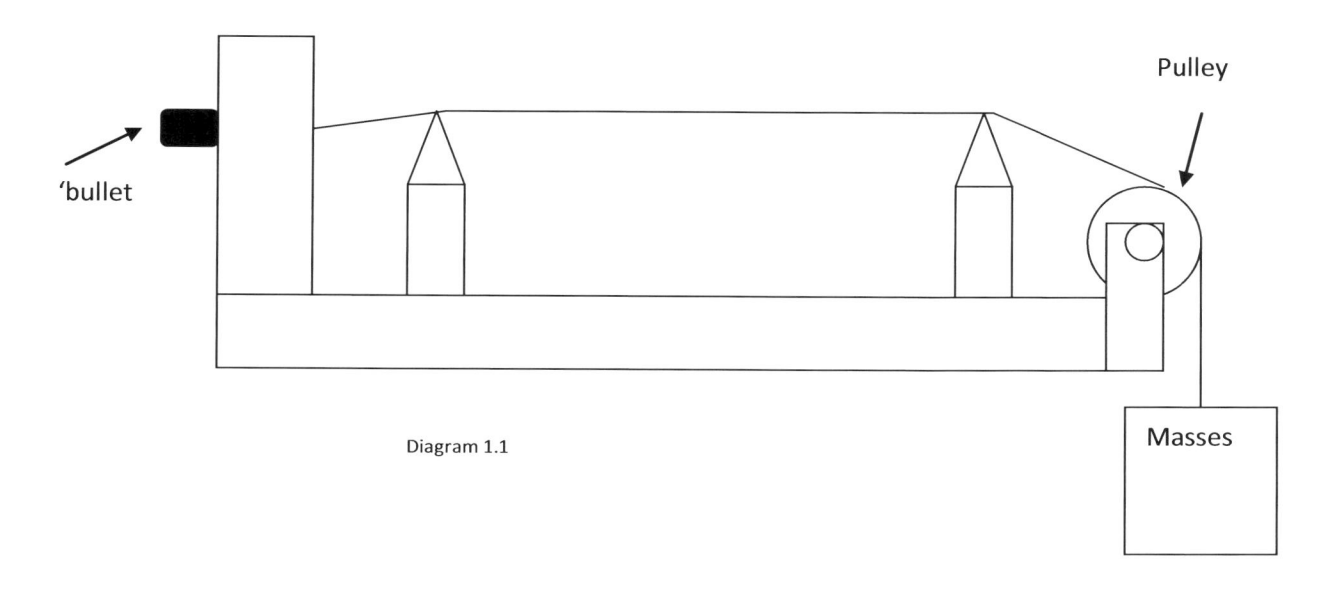

'bullet

Pulley

Masses

Diagram 1.1

Here is an example of part of a good write-up for the procedure. It contains a labelled digital photograph.

EXAMPLE BEGINS HERE

Experiment 1: Driven Bicycle Wheel

The object of this procedure was to investigate the angular acceleration α, of a bike wheel, with the applied Torque. In order to look at this relationship the wheel was accelerated from rest with a constant torque and repeated with 8 different torques at set increments.

The apparatus used in this investigation was a bike wheel, wall mounted on axles so that it could rotate vertically as in a real bike. In order to wall mount the wheel, a frame was constructed, consisting of 2 sections of 2" x 4" wood and an MDF sheet. The wheel was attached to the frame using quick release axles through 2 angle brackets either side of the wheel. The frame was attached to the wall by adding two U brackets which allowed the frame to hook onto the white board rail.

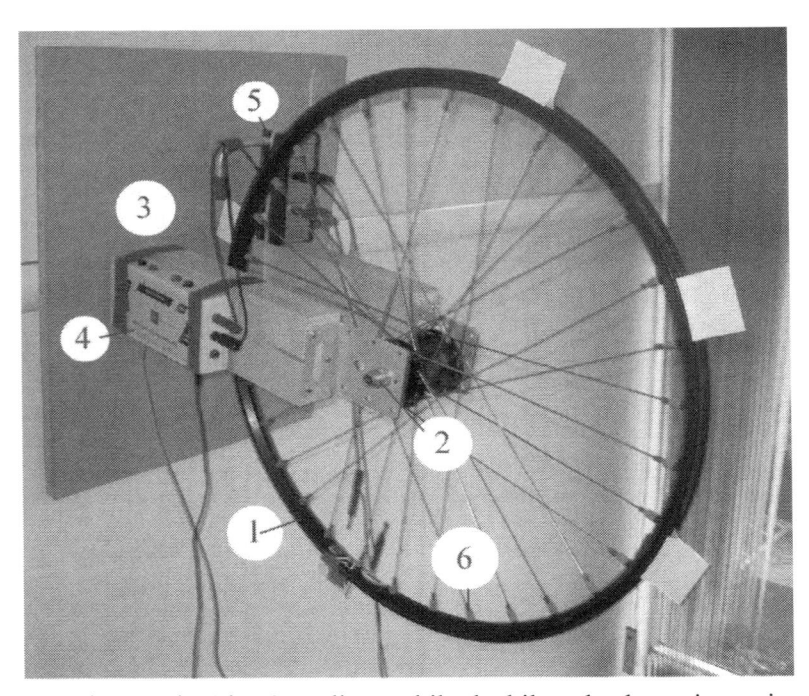

	1 Wheel
	2 Axle skewer
	3 Wall mounting frame
	4 ALBA Interface (connected to PC)
	5 Light gate
	6 Nipples

In order to take 'time' readings while the bike wheel was in motion, a light gate was mounted, set up so that as the spoke nipples passed the light gate they cut the light beam. This obstruction of the light beam would then send out a signal to an ALBA interface connected to a computer which logged the time each spoke nipple took to pass the light gate. It was decided to use the spoke nipples as timing marks as they were equally spaced around the rim and as there were so many (32) they would provide a more detailed and accurate account of the wheel's motion.

By measuring the diameter of the nipples (**s**) and the distance from the centre of the wheel to where the nipple cuts the light gate (**r**) the angular displacement could be calculated by using the equation:

$$\theta = s/r.$$

Giving a subtended angle of θ = 0.0077/0.285 = 0.027 radians per spoke. This, combined with the time taken to for the nipple to pass the light gate, allows the angular velocity at that point to be calculated with the equation:

$$\omega = \theta/t.$$

The nipple width **s** was measured using a screw gauge micrometer to give increased accuracy and **r** was measured using a metre stick.

In order to accelerate the wheel with a constant torque, a string was attached to the wheel on a hook in the wheel rim. The string was then wound once around the wheel rim and a mass was attached to a loop in the end of the string. The mass could then be released allowing it to accelerate due to its own weight, thus giving a constant torque. To provide 8 different torques 8 different masses were used from 10-80g at 10g increments.

The magnitude of the torque is given by:

T=Fr

and since **F = W = mg**, this becomes:

T=mgr.

Since the string sat in the concave wheel rim, it was difficult to find the distance from the axis of rotation to where the torque acted (**r**). Therefore to measure this, a string was wrapped round the wheel in the rim and its length was measured. From this, the radius was calculated using r = circumference/2π.

Initial trials of this experiment showed that the graph of θ against ω^2 did not give a smooth line as expected with constant torque. A hypothesis as to the cause of this was that the nipples were not of a constant diameter (as they had flats so the spokes could be tensioned with a spanner). To investigate whether or not this was the cause, the nipples were widened by sheathing them in a rubber hose to give a constant diameter (the nipples were all checked individually to be sure of this). This measure improved the graph. Increasing the diameter increased the time it took for each spoke nipple to pass the light gate, reducing the fractional timing uncertainty. It also increased the angle subtended by the spoke θ, so uncertainties in θ were relatively smaller.

Another step taken to improve accuracy was to carry out repeat readings. For each Torque increment, five sets of timing results were taken. The average of these five times was then calculated and angular velocities computed from these mean values. This also had a noticeable effect in smoothing the graph of θ against ω^2.

EXAMPLE ENDS HERE

Section 3   Results

> - **Data sufficient and relevant** to the purpose of the Investigation.
> - **Uncertainties** in individual and final results.
> - Appropriate **analysis** of data, eg. graphs, calculations.

This is where you produce all the data to back up your conclusions. Include enough of it to satisfy the most persistent interrogator. If you've been using your daybook properly, it should just be a case of transferring the material into a neat form. Here is an example from a capacitor Investigation:

>>>>>>>>>   EXAMPLE BEGINS HERE   <<<<<<<<<

Circular Plates

1. Area of Plates:

The diameter of the circular surface of each plate was measured, and an area calculated. The diameter of the insulator hole was also measured and an area due to this hole calculated and then subtracted from the circle area, giving the area of the plate. Each measurement was done several times and an average was taken. The process was carried out for both plates, they were compared and the area of overlap was taken as the Area of the plates.

Plate 1 -

Diameter = 325 mm
326 mm
324 mm
d_1, average = 325 mm r_1 = 162.5 mm, error ± 1 mm

Diameter of
insulator hole = 13 mm
13 mm
13 mm
d_2, average = 13 mm r_2 = 6.5 mm, error ± 1 mm

Area of Plate = Area of circle − Area of insulator hole
$= \pi r_1^2 - \pi r_2^2$
$= (\pi \times 162.5^2) - (\pi \times 6.5^2)$
$= 26406.25\pi - 42.25\pi$
$= 82957.68 - 132.7323$
$= 82\,825 \text{ mm}^2$
$= \underline{0.082825} \text{ m}^2$

Error in d_1 = $\frac{1}{325} \times 100$ = 0.31 %

Error in d_2 = $\frac{1}{13} \times 100$ = 7.69 %

Error in A = $\sqrt{\left(\frac{1}{325}\right)^2 + \left(\frac{1}{13}\right)^2}$

= $\sqrt{5.927 \times 10^{-3}}$
= 0.07698
= <u>7.7</u> % to 1 d.p.

Plate 1 Area
$$(0.0828 \pm 0.0064) \, m^2$$

Plate 2 -

Diameter = 326 mm
326 mm
<u>325 mm</u>
d_1, average = <u>325.67</u> mm r_1 = 162.83 mm, error ± 1mm

Diameter of
insulator hole = 13 mm
13 mm
<u>13 mm</u>
d_2, average = <u>13 mm</u> r_2 = 6.5 mm, error ± 1 mm

Area of Plate = Area of circle – Area of insulator hole
= $\pi r_1^2 - \pi r_2^2$
= $(\pi \times 162.83^2) - (\pi \times 6.5^2)$
= 83294.959 – 132.7323
= <u>0.083162</u> m^2

Error in d_1 = $\frac{1}{325.67} \times 100$ = 0.31 %

Error in d_2 = $\frac{1}{13} \times 100$ = 7.69 %

Error in A = $\sqrt{\left(\frac{1}{325.67}\right)^2 + \left(\frac{1}{13}\right)^2}$

= <u>7.7</u> % to 1 d.p.

Plate 2 Area
$$(0.0832 \pm 0.0064) \, m^2$$

Area of Plates, A - area of overlap, round the answer to 3 decimal places;

$$(0.083 \pm 0.006) \, m^2$$

Distance between plates:

To measure the plate separation three triangular perspex spacers were used. These were placed between the plates such that the distance separating the plates could be set. The screw on the insulator rod was then tightened to hold the plates in place and the spacers were then knocked out. The height of the spacers had to be measured, several times so an average could be taken.

Spacer 1 -   Height Measurements
               1 = 2650 µm
               2 = 2660 µm
               3 = 2660 µm
         Average = 2656.67 µm

Spacer 2 -   Height Measurements
               1 = 3030 µm
               2 = 3070 µm
               3 = 3060 µm
         Average = 3053.33 µm

Spacer 3 -   Height Measurements
               1 = 2980 µm
               2 = 2970 µm
               3 = 2970 µm
         Average = 2973.3 µm

An overall average was taken for the three spacers' averages and a large uncertainty was taken, this gave the distance between the plates

Average = (2656.67 + 3053.33 + 2973.33) / 3
        = 2894.44 µm

Uncertainty = Range / 3
             = (3053.3 − 2656.67) / 3
             = 132.22 µm

The micrometer can only read accurately to 10 µm so this reading error was added to the error in the distance between plates.

Distance between plates, d:
                    (2894 ± 140) µm

EXAMPLE ENDS HERE

Section 4   Discussion / Conclusions

> - **Conclusion**(s) relevant to the purpose of the Investigation.
> - **Evaluation** of experimental procedures to include, **as appropriate**, comments on:
>   — accuracy of experimental measurements
>   — adequacy of replication
>   — adequacy of sampling
>   — adequacy of control of variables
>   — limitations of equipment
>   — reliability of methods
>   — sources of errors and uncertainties
> - Coherent **discussion** of overall conclusion(s) and critical evaluation of the Investigation **as a whole**, to include comment, **as appropriate**, on:
>   — problems arising and how they were overcome
>   — modifications to procedures
>   — significance/interpretation of findings
>   — suggestions for further improvements to procedures
>   — suggestions for further work

The size of the box above would suggest that you could be in for a lot of writing. The trouble is that you often find there isn't much to write about. Physicists are trained to cut to the chase, so it goes against the grain to spew up a load of waffle; just don't be reluctant to state the obvious (it'll fill up that blank page a bit and push-start the linguistic flow).

The conclusion, only worth one mark, can be fairly short and direct, even if it's negative. Keep it sharp and snappy; make it sound scientific. Here is an example of a conclusion taken from an aerofoil / wind-tunnel project. This type of investigation can be a bit short of data, so authors are often struggling for conclusions. This student makes a decent attempt at it (though his sentences are too long).

Section 7: Conclusion

Aerofoil 2 was by far the leader in terms of lift force generation and efficiency, it did, however, appear to stall very easily at angles beyond around 35°. All of these properties could be directly attributed to the effect that the sharpened leading edge has on postponing flow separation and decreasing turbulence up until 35°, after this the angle through which the air flow had to turn was too great. As far as drag force generation went, aerofoil 3's was generally lower that the other four. This was most likely due to the smooth shape allowing the air to flow without turning through too sharp an angle, which causes increased turbulence and induces flow separation. Aerofoils 1 and 3 have similarly stable lift force graphs, only aerofoil 1's has a greater amplitude. Again this could have been attributed to the path that the air was forced to follow, as the air was not forced to turn through too sharp an angle the flow remained relatively stable. Aerofoil 4 was a great disappointment, instead of improving on the design of aerofoil 3 I only succeeded in making it worse by forcing the air to turn through too much of an angle. The alternative design for aerofoil 4 was only added after I had gathered all of my data and graphed it all.

In conclusion Aerofoil 2, was by far the most efficient and highly versatile. Aerofoil 1, with its high lift / high drag combination, would have been suitable for low speed flight at most angles of incidence, if the maximum angle of incidence required was less than 35° the aerofoil 2 would be used instead as it was more efficient. Aerofoil 3, with it's low lift / low drag combination, would have been suitable for high speed flight where keeping the drag force as low as possible is essential. Aerofoil 4 could do nothing that one of the other aerofoils could not do better. The experimental design of aerofoil 2 functioned more than satisfactorily with the sharpened leading edge reducing drag and increasing lift to a more than reasonable extent. The experimental design of aerofoil 4, however, was a complete failure with reduced lift and increased drag.

All in all, a good effort. This next example combines the conclusion with the evaluation. It's about monitoring motion using Lissajous figures.

>>>>>>>>>> EXAMPLE BEGINS HERE <<<<<<<<<<

Discussion

The conclusions gathered from this investigation are as follows:

Overall, the experiment was a success, and has shown that the phenomenon of Lissajou's Figures can be used as a tool to measure the acceleration of a moving object. Using Lissajou's figures to find certain points along a horizontal plane appears to be an extremely accurate way of achieving this aim.

It would have been easier to use a logic mechanism, rather than the method of monitoring the oscilloscope screen, to find the points when both the original signal and the observed signal are high; this would allow the phenomenon to be observed directly which would afford an even greater degree of accuracy. It would also simplify the experiment by removing several components, such as the ALBA hardware and the oscilloscope. The reason that this was not done was due to time pressure, as it was not possible in the remaining time to design and construct a logic circuit capable of carrying out this task.

To gain greater control over the motion of the vehicle, it would have been necessary to reduce the effect of friction from the dragged cable from the microphone. There was no obvious way to do this, other than sending the signal from the microphone "wirelessly"; this may have been achieved by encoding the information into a digital signal and sending it to a receiver via a radio or microwave transmitter, or alternatively a laser or LED. However, due to limited resources, none of these options were directly available.

The masked cable, used to carry the signal from the microphone, greatly reduced electromagnetic interference; this was aided by covering the positive terminal in tinfoil, which was held at ground by connecting it to the 0V terminal. If the resources were available, it would have greatly improved the signal from the photodiode by applying the same technique, and fully masking the cable. [See diagram]

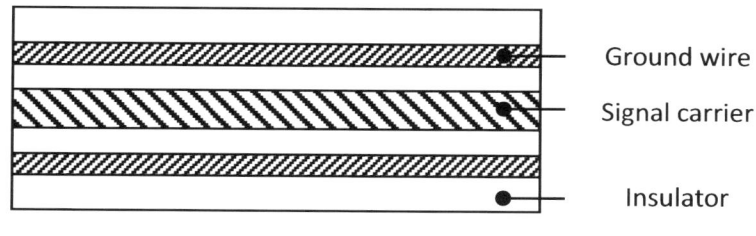

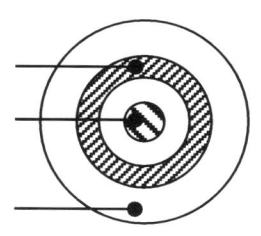

Cross section of cable (side)   Cross section (front)

Ground wire
Signal carrier
Insulator

>>>>>>>>>> EXAMPLE ENDS HERE <<<<<<<<<<

This next example comes from a study of radon gas in buildings. The TASTRAK material mentioned is a polymer sensitive to the passage of α-particles.

Evaluation of Methods

The results gained from both methods used to measure the radon levels were quite promising as they were comparisable in most cases.

A gas sampling mean result of 197.33 had a corresponding TASTRAK mean reading of 55 which gave a possible radon concentration of 48.583 ± 4.583 Bq m^{-3}.

In another instance gas sampling gave 238.33 with a TASTRAK reading of 145.5 giving a possible radon concentration of 128.525 ± 12.125 Bq m^{-3}.

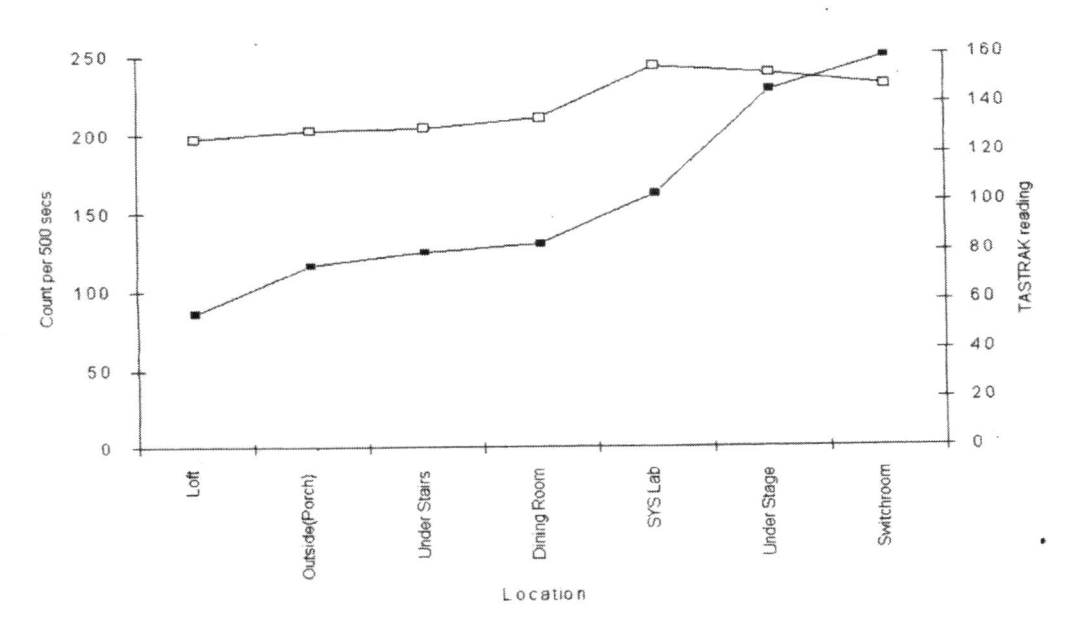

The graph shows the two sets of readings and gives a good indication that both methods are getting similar results. The unusually high readings for the gas sampling in the Sys lab and for the TASTRAK plates in the switchroom are attributed to the special conditions occurring there. Otherwise the two graphs contain similar trends.

Special Cases for Gas Sampling

The readings taken in the SYS lab show an unusually large amount of counts taken but the TASTRAK plates show a much lower level of radon. This is probably due to a high background count which was picked up by the Geiger Mueller tube, since TASKTRAK plate is shielded from this interference.

Special Cases for the TASTRAK plates

In the Switchroom, several plates were placed in different locations to see if the radon gas was concentrated in any places and I found that TASTRAK counts were higher around power cables to the tune of about 220, 50 cm away from the power cable to 154, 150 cm away from the cable.

Section 5   Presentation

> - *The Report must include a title, table of contents and pages must be numbered.*
> - *The Report must be clear, concise and readable.*
> - *The sequence and development of ideas must be logical.*
> - *There must be sufficient detail to allow the Investigation to be repeated.*
> - *The length of the text should be around 2500-4500 words, excluding tables, figures, graphs.*
> - *References must be sufficient, relevant and specific.*

Separating the objective from the subjective is the most important process in science. To a physicist, a title like 'presentation' is liable to conjure up thoughts of shallow marketing executives in front of flipcharts, or advertisements designed to empty his wallet. It seems to go against the grain of everything he's been taught; the years spent cutting through the crap to be replaced by capitulation to the men in silky suits; or that invitation to mark up the glossy but shoddy piece of work in his hands. Inviting a **real** physicist to give time to a presentation is like asking the Archbishop of Canterbury to believe that God doesn't exist.

I'm sure the SQA didn't intend awarding marks to students who can make a silk purse out of a sow's ear, but this is an opportunity to gain marks without knowing any physics. References must be in a specific format? Marks awarded for page numbering? The summary must be the first item in the introduction? It's easy to be critical of this approach, and it's also easy to take a stand and reject it. But if you do, then you just shoot yourself in the foot. My advice is to be like Leibnitz and Kant: keep your real thoughts private. In public, put some of the gloss back in and sell yourself.

Use a Computer to Write the Report. There's no penalty for producing a handwritten Report, though very few students do it this way. Using your own mobile device at school and home is ideal since you don't have to stop when the bell rings (or the janitor chucks you out), and it also means you're not limited to the software on the school machines. You can use Flash memory to exchange data between a school laptop (where you might record the data), and your own laptop (where you analyse it). Using your own computer at home should relieve you of some distractions, but bring in a whole host of others!

Most students will be familiar with programs like Word and Excel. For those who aren't, here are a few hints and tips with special relevance to the writing of a Report.

Firstly, get yourself organised. Create a new folder to keep all your project work inside, with subfolders keeping separate things like graphics, photographs, correspondence, spreadsheets etc. Give some thought to naming files / worksheets if, for example, you have twenty different Excel datasets. When using Word:

- Set the margins for the Page Layout. For A4 try 2cm in on all sides.
- Set a line spacing of 1.5.
- Choose a standard, readable font at a decent size (12pt – 14pt) rather than an ornate one (my own personal dislike is comic sans).
- Minimise the number of fonts used. Ideally, just use one font throughout the document, using different sizes (but not too many), bold or italic for titles / highlighting / drawing attention.
- Space out the material without giving the impression that you hate trees. Print single side.
- Remember that formatting is paragraph based and that the end of a paragraph is defined by use of the <enter> (or <return>) key. Try to get proficient at formatting.
- Unless you're proficient, use page/chapter breaks sparingly (and find out how to remove them).
- Keep page numbering simple (do you really need a 'header'?)
- If you create a vector drawing in Word, make sure you group the elements together when finished.
- Get proficient at positioning graphics. Avoid the 'inline' placing. Use 'square' for smaller graphic objects, or 'top and bottom' for graphics spanning the width of the page.
- Remember that graphic positions are attached to a particular paragraph by default (which is why they sometimes annoyingly slam up against the top of the page).
- Crop / manipulate / add captions to photographs using custom software like Photoshop Elements (if available) before pasting into your document. What you see on the screen won't quite be the same as what's printed.
- Consider using labels for graphs / tables (like 'graph 1' etc) if you've lots of them.

Writing reports using IT like Word and Excel, or manipulating and creating graphics using commercial software like Adobe Photoshop / Illustrator, is a very useful skill for the future. Take this opportunity to get familiar with as much of it as you can get access to, and use your report as practise for the real world. But at the end of the day, it is a physics report and must pass this final hurdle,

is your Report clear and intelligible to another physicist unfamiliar with your work?

Finally, don't forget to give references and acknowledgments. Keep the latter simple (including your favourite band doesn't set a good impression). An example of an SQA approved reference is:

Resnick R & Halliday D, (1966), Physics Part 1, John Wiley & Sons, London

In the Lab 5

Safety

A memorable (often risky) experiment is what turns on many youngsters to science in the first place. Once they discover what science is all about, they can sustain their interest without the need for a constant diet of explosions and bangs (or, at least, some of us can). But, towards the end of the last century, new safety regulations were introduced which had an effect like a body blow on teachers. The old regime of excitement and mishaps was replaced by paperwork and dull experiments. The Health and Safety Executive (HSE), the body in charge of its implementation, insisted that they weren't there to spoil the party; but to teachers, it was like a directive from Joseph Stalin. The publicity was so bad that the HSE had to fight back with their 'myth-buster' calendars; the one on the right is an example.

Given that setting, where does it leave the Advanced Higher Physics student and teacher? The nature of the course is such that you will be working unsupervised for most of the time on experiments which may be unfamiliar. The advice which HSE gives to cover all occasions is its 'Five steps to risk assessment'. The web address is: http://www.hse.gov.uk/pubns/indg163.pdf

Step 1	**Identify the hazards**
Step 2	**Decide who might be harmed and how**
Step 3	**Evaluate the risks and decide on precautions**
Step 4	**Record your findings and implement them**
Step 5	**Review your assessment and update if necessary**

For school experiments, the teacher usually goes through the procedure himself, but at Advanced Higher it's a good idea to go through it jointly with the student (though clearly responsibility lies with the school/teacher). The HSE provide a general, risk assessment proforma (reproduced on the next page) which takes you through the process. Many physics departments will have their own version of it. Remember that much of what you do will be low hazard / low risk. If your experiment involves a medium hazard, your aim is to make it low risk; that's the point of the Health & Safety exercise. As the HSE puts it,

Our approach is to seek a balance between the unachievable aim of absolute safety and the kind of poor management of risks that damages lives and the economy.

Company name:

Date of risk assessment:

Step 1 What are the hazards?	Step 2 Who might be harmed and how?	Step 3 What are you already doing?	Step 4 What further action is necessary?	Step 4 How will you put the assessment into action?
Spot hazards by: ■ walking around your workplace; ■ asking your employees what they think; ■ visiting the *Your industry* areas of the HSE website or calling HSE Infoline; ■ calling the Workplace Health Connect Adviceline or visiting their website; ■ checking manufacturers' instructions; ■ contacting your trade association. Don't forget long-term health hazards.	Identify groups of people. Remember: ■ some workers have particular needs; ■ people who may not be in the workplace all the time; ■ members of the public; ■ if you share your workplace think about how your work affects others present. Say how the hazard could cause harm.	List what is already in place to reduce the likelihood of harm or make any harm less serious.	You need to make sure that you have reduced risks 'so far as is reasonably practicable'. An easy way of doing this is to compare what you are already doing with good practice. If there is a difference, list what needs to be done.	Remember to prioritise. Deal with those hazards that are high-risk and have serious consequences first. Action Action Done by whom by when

Step 5 Review date:

■ Review your assessment to make sure you are still improving, or at least not sliding back.
■ If there is a significant change in your workplace, remember to check your risk assessment and, where necessary, amend it.

Home from home; the Physics Lab

Most schools will have a lab set aside for sixth year physics students. It's often the smallest room in the department and tends to accumulate the bits n' pieces that the teachers don't use. You could be lucky and have lots of bright bench space, but you're more likely to be sharing not many square metres of bench with the rest of your group. Get organised. As soon as you get the go-ahead from your teacher, 'acquire' some bench space that suits your project. Some of your fellow students will require more space than others; deciding who goes where won't be a problem. Look for storage space where you can keep things out of harm's way. You may be able to set up your apparatus and leave it between sessions ready for a quick start next time; though in some schools, space may be so tight that a teacher will require use of the lab to demonstrate an experiment to junior classes. Moving or dismantling equipment can feel like a major hassle but it's not; it's minor hassle and you'll take things in your stride since it won't happen often. Discuss what equipment you'll need with your teacher. Ascertain if you can have sole use throughout the duration of the project or if the teacher will need it at certain times in the session. Will you need something specially built or adapted by a technician? This can be a tricky area. Technicians vary greatly from school to school; some of them are highly skilled individuals keen to help, others are only capable of pushing first year trolleys between labs. It would be a bad mistake to treat them as 'mere' technicians. Be pleasant, and don't be reluctant to say 'thank you'; otherwise leave the teacher to liaise on your behalf.

Once you've got your bench space sorted out, you'll need to set up the equipment. If possible, clear everything else away and give yourself room. Don't have piles of wires mixed with odd bits of junk pushed to the back; tidy it away, if necessary asking your teacher where to put it. Where do you leave your bag and jacket? On the floor up against the bench is not a good idea; try to hang them up out of the way but consider the security aspect if they contain valuables. Has your workspace got the required number of mains sockets? Will you need a 4-way multi-adapter? Do you need a gas tap? If so, where is the main on/off valve? Does it need a key? Will you need access to a sink nearby? And will you need that important item – the G-clamp? This can very useful: from clamping down the stand of a pendulum experiment to stopping a vibration generator making the string slack in a standing wave. Usually they are in short supply and you have to share with other students (borrowing one from your parents is a good move).When you've finished your experimental session for that day, can you leave the equipment set-up? Are you required to return certain pieces of apparatus (like a laser) after each session? Should you leave weights hanging on wires? Is it wise to leave hot objects to cool down unattended? Does anything need to be cleaned (it's always better to do it today rather than leaving it until tomorrow). Should the last student lock the lab door? The Physics Department will have a set of Laptop computers for pupil use. If you need the use of one for your project, familiarise yourself with the procedure for borrowing / returning. If you leave the lab with the Laptop connected to your equipment, do you need to lock the door? If you use your own laptop, has it got the correct connector type to interface with the experiment?

Keep your teacher informed of the progress of your experiments on a weekly basis. There is a type of personality which bottles up problems, either through natural shyness or unwillingness to admit 'failure'.

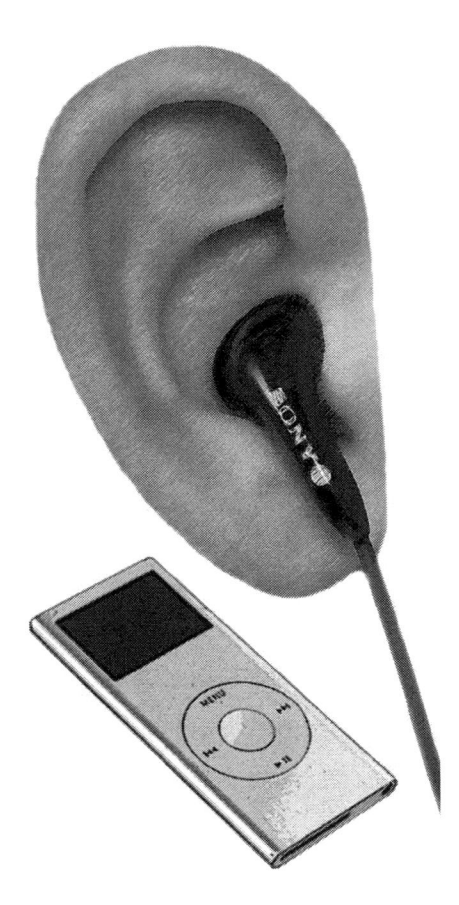

Don't let this happen to you. Your teacher may not have all the answers, but discussing the problem and bouncing ideas off each other is part of the scientific method. Take the opportunity to raise any safety issues you may have, from cracked sockets to chemical odours, and report any equipment which isn't working properly. Write down a synopsis of these discussions in your Daybook (and remember to use the Daybook every session for routine things like noting down 'live' data).

The things I've covered so far have been fairly straightforward. Here are a few of the trickier issues where a student (and the teacher) might feel pressured or restricted. The first one is dress. Bits of clothes which hang down or stick out can be a mild nuisance if they continually need to be pushed out of the way (why give yourself the hassle), or a safety issue if they get caught in moving equipment. Dangling metal 'bijou' may provide a conducting path and be an electrical hazard. Why not just keep things simple? You're bright enough to do AH physics so you don't need to make a statement through your clothes.

The next one is music. How this is treated varies from school to school and teacher to teacher. MP3 players have become part of a student's anatomy, and combined with the more relaxed atmosphere of sixth year it's natural to assume that you'll be able to use them in the lab. Here are some points to consider. First of all, ask your teacher if it's okay to use an MP3 player in the lab. Some will say yes, some will say no. But it's bad manners just to assume that it's okay. There are good safety reasons for refusal based on your impaired ability to hear external sounds. Whatever decision your teacher makes, you should accept it gracefully. More contentious is the sound system with speakers entertaining the whole group. Selecting music which everyone can stand isn't the problem; neither is the safety issue mentioned above (unless it's too loud). The problem is that it's in a school and there are junior pupils around. Schools try hard to create the right 'ethos' (they even get assessed on it by HMI), and the dulcet tones of the latest popular beat combo (I'm being kind) floating out from the sixth year lab might give passing junior pupils the impression that schools are social venues rather than places of learning. A touring deputy rector might also form a poor opinion of your teacher. You don't have to behave like the 'Famous Five' but you've got to try to be a positive role model for younger pupils. On the topic of what sounds float out of the lab door, you might need to consider your language. An experiment which doesn't behave as expected needs a bit of thought, not an expletive. If the door is open, your description of last night's activities, choice words and all, might reach inappropriate ears. And let's be 'up front' here; it's not the place to conduct a relationship. How about eating food or drinking liquids in the lab? Always ask your teacher. They will probably be happy enough if you have a bottle of water, but less so about solid foods (there are regulations concerning the consumption of food where chemicals etc are stored, so don't make a fuss).

The common theme in all the above is *responsibility*. Give it a try.

Do you believe in Hobgoblins?

Here's the secret to doing experiments. You'll be testing inanimate lumps of matter; things like light bulbs, beakers of water, lumps of semiconductor, dollops of plasticene, metal rods, sheets of polystyrene, beams of electrons and rolls of wire. They all have something in common; they don't have a brain, they can't think. That light bulb you're about to use doesn't think to itself,

> .... *'look, he's coming back to do an experiment on me, let's play dead today'*.........

If it's in working order, **it will operate**. If it's not, **it won't**. I can guarantee it. That's the way our universe is built. If it wasn't like that, I would give up Physics and take up Biology. Contrary to popular belief, there are no hobgoblins behind every boulder waiting to trip you up.

Physics experiments **always** work. They may not work as you expected, you might even be disappointed, but they always work. Try driving off in a car with no petrol in the tank, and it won't start. There's a reason for it not working, and in this case the solution is simple; put petrol in the tank. It's the same with physics experiments in the laboratory. You find out what the problem is, and then try to fix it. Don't think it's just **YOU**. The fact that you're doing this course means that you're bright enough to meet the challenge.

The basis of experimental physics is measurement. The simplest, may require the distance between two points. Quite straightforward, if all you have to do is place a 30 cm ruler over the object and read off the result; though surprisingly problematic over longer and shorter distances, or if the ruler can't access the end points of the object, or if it's along a curve. Timing can also present problems; not so much over the accuracy of the instrument, as defining the start and finish events. Unlike distances, longer means better; but like distances, shorter means trouble.

With distance and time, what you see is what you get (unless you're into Special Relativity). There are no hidden nasties lying in wait; unlike electrical measurements, where you may have to consider the method of measurement. Sometimes, there's more to it than simply placing multimeters in the right position. Even more difficult is the measurement of light; dependent upon the nature of the detector and, for white light, complicated by its range of wavelengths. And there are others . . . . .

Experimental physics may not be as sexy as dark energy or quantum gravity, but it has its fascinations and challenges for those with an enquiring mind. In the next section, we look into some of the routine measurements carried out in a physics investigation, and consider the strengths and weaknesses of the instruments used to perform them. Leave any thoughts of hobgoblins at the door on your way in.

Instruments - Measuring Distances

This is the bread and butter of experimental physics and the most convenient instrument is the 30 cm ruler. The traditional wood construction has been largely replaced by steel and plastic, denying teachers the pleasure of picking out the rogue ruler from the class set. When introducing the topic of systematic uncertainty, the teacher would pass a sheet of A4 paper around the class, asking them to measure its width to the nearest millimetre, but sneakily switching rulers half way. One half of the class measured the width of the paper with the rogue ruler while the other half measured it with an accurate ruler. After displaying the results on the board, noticing the discrepancy, and looking puzzled for a suitable interval, the teacher could triumphantly announce the ruse and drive home his point.

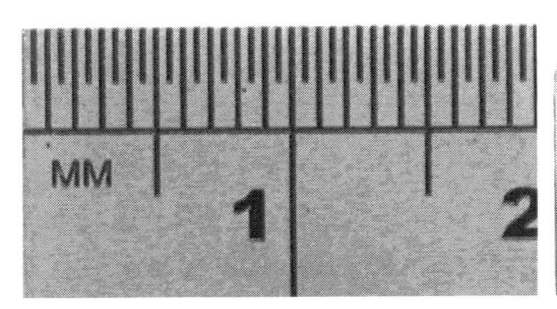

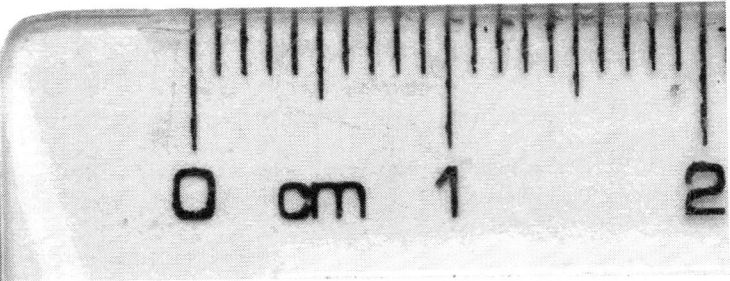

The photograph above shows close up views of the ends of a 30 cm aluminium ruler (on the left) and a 30 cm plastic ruler. Both rulers have only had light use so the edges and corners are still fairly new. The scales of plastic and most wooden rulers don't start at the ends to prevent errors due to wear, though metal rulers usually have the zero at the edge. The steel ruler (bottom left photo – notice the end error), beloved of technical departments and engineers, can be difficult to read due to the lack of contrast and the dull shine of the material, often making its half millimetre markings rather academic without a hand lens. Many rulers are designed with a bevelled edge along the scale to minimise parallax error. Are metal rulers better than wood or plastic? Study this photograph and judge for yourself (the bottom right ruler is made of aluminium).

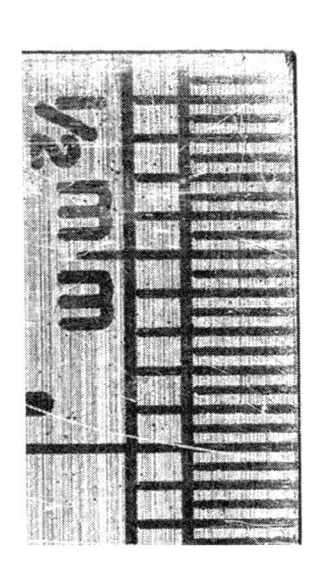

Longer distances require a metrestick. They don't have a bevelled edge (which is at least 3mm thick), so must be read with your eye vertically above the reading. Nor do they have extra material at each end; the scale starts and finishes on the ends. Here's what the end of a metrestick can look like after a few years.

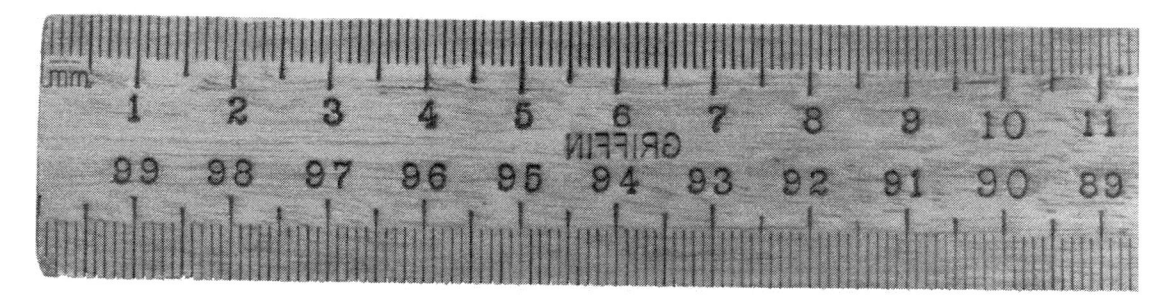

Notice the rounding on the corners (especially the bottom one), though this wouldn't be a problem if you use the middle of the metrestick to align your measurement (since the middles' are less likely to be abraded). The lack of deadwood at the ends also allows metresticks to be stacked for measuring longer lengths.

Longer than ordinary rulers, and constructed from a natural material which can change its size with moisture and temperature, a measurement with a metrestick should have a greater absolute uncertainty. If you have a number of metresticks, compare them with each other by lining up the zero of one with the 100cm mark of the other. Now check the alignment at the other end. Repeat for all the metresticks and discard any rogues. Pick one from the consistent batch. There's still the possibility of a systematic error, but there's nothing you can do about it unless you have access to your Local Authority standard metre.

Longer distances still, require instruments like tape measures and trundle wheels; or electronic gadgets based on laser light or ultrasound. The laser device (centre) measures distances up to about 100m with an accuracy of about ± 5mm, while the ultrasonic device measures out to about 15m with an accuracy of ± 0.5%.

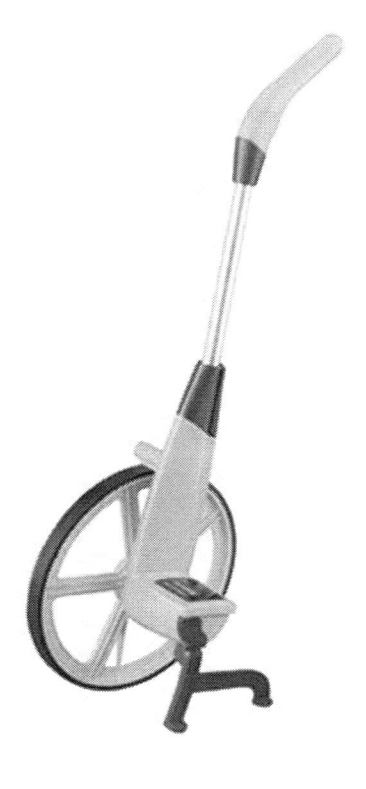

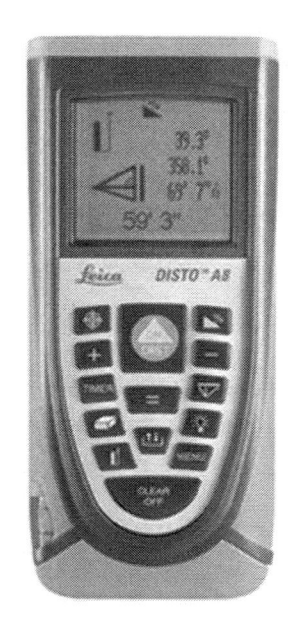

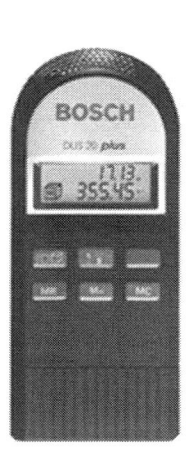

Smaller distances tend to be a bit trickier to measure and require more expensive equipment, of which the prime example is the vernier microscope. The 'vernier' word refers to a cunning method for gaining another decimal point or two of accuracy. The photograph shows a travelling (since you can move the view) vernier (since it has a vernier scale) microscope (since you need to magnify the small object you're measuring).

You can spot two eyepieces for looking through. The bigger one on the left is for looking at the object to be measured (placed on the ground glass illuminated screen; it's often a slide). The smaller one at the back right is for reading the vernier scale (which is just below and to its right).

Here's how it works. There are two scales, a fixed main scale from which you read the measurement (you can see it running along the full length of the back of the travelling microscope in the pic), and a small sliding scale (the vernier scale). Refer to the drawing below. The vernier scale marked from zero to ten is exactly nine tenths of the distance from zero to 1cm on the main scale. That's really sneaky, because if you line up the zeroes on both scales then move the vernier scale until the ten lines up with the 1cm mark, you must have moved it by exactly 1mm on the main scale. If you'd moved it by only ½mm, the 5 of the vernier scale would line up with 5mm mark on the main scale. Let's take the reading on the diagram. Imagine the zero on the vernier scale lined up with the 5mm mark on the main scale. Move the vernier scale to the right to the position shown in the diagram. The marks on both scales line-up along 'AA'; which is three tenths of the way along the vernier scale. So the zero of the vernier scale is three tenths of the way between the 5mm and 6mm marks on the main scale (at point 'P'). This gives a reading of 0.53mm. Check the one below gives 3.268cm.

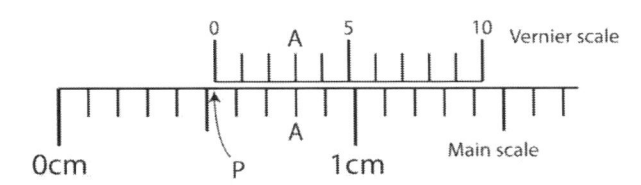

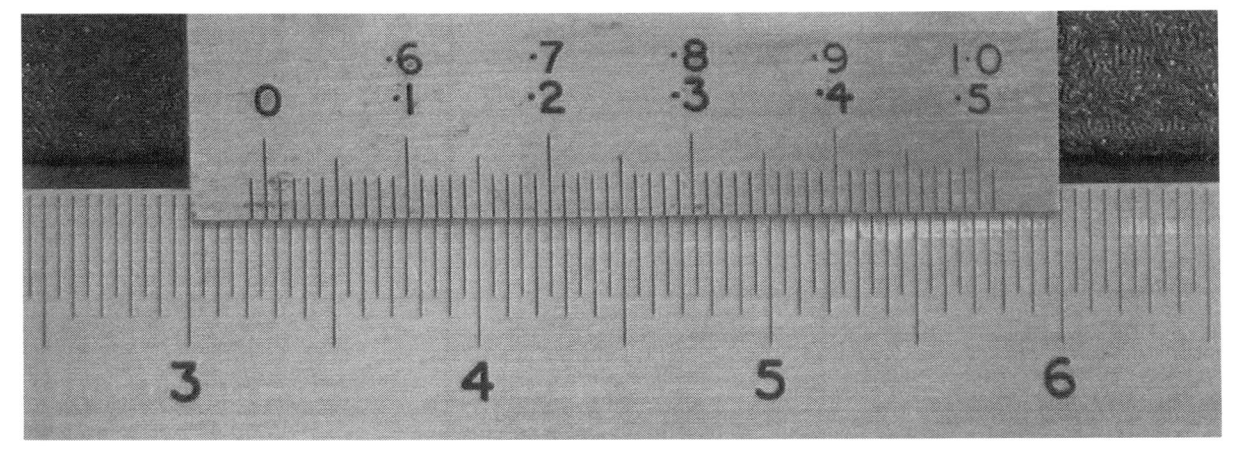

Another common device for measuring short distances with precision is the **micrometer screw gauge**. The one in the diagram above is shown with its jaws shut (at 'A'). Notice that the zero is in the correct place on the scale; you should check that it does this. The barrel (which is the right half in light grey) has two roughened surfaces. The larger of the two is the coarse adjustment screw for moving the jaws quickly, but, you mustn't use it to tighten the jaws against the object to be measured (it puts too much of a strain on the internal mechanism). Instead, use the small, final adjustment screw on the end. The inset shows a gauge with a vernier scale inscribed around the inner barrel; making it a vernier micrometer screw gauge (it reads 0.5783cm).

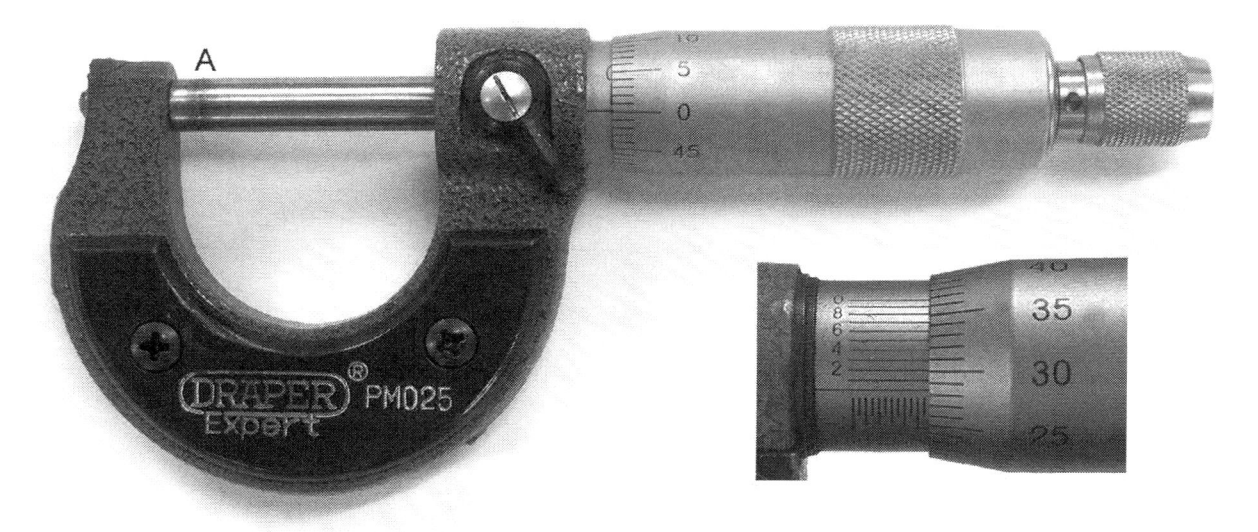

A **calliper** is a two-legged instrument for measuring distances. Usually it's a parallel sided object you're measuring such as a cuboid or a cylinder, or a sphere; but it doesn't have to be. The one shown below is an uncommon version called a dial calliper; more often that not, you'll meet a vernier calliper. The two big pointy callipers sticking out the bottom are for going around the outside of the object (like the outside diameter of a pipe), and the two little ones at the top fit inside the object (like measuring the inside diameter of the pipe).

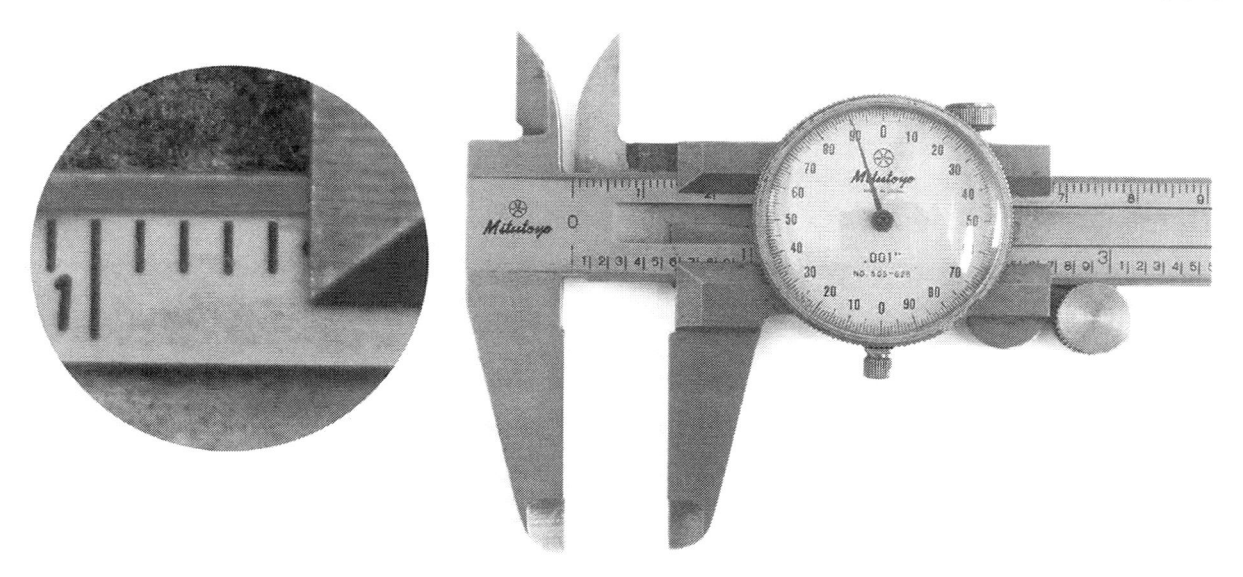

Micrometers and callipers are precision instruments. You can appreciate the care and attention to detail that has gone into their design and manufacture. Look after them.

Measuring Times

Using a hand-timer will be appropriate for many experiments (there's no point overcomplicating things after all), but for Advanced Higher Physics you should be looking for methods which remove the human element.

Mobile phones and mp3 players are often the hand-timer's instrument of choice. They've replaced the traditional stopwatch even though the buttons are more fiddly to operate. They display times to one hundredth of a second and calibration uncertainties are insignificant. One thing I would ask you not to do is use your mobile phone or mp3 player without permission. Teachers try to relax the rules for students in sixth year, but schools also have policies on their use, and it would be rather selfish to put your teacher under pressure through special pleading. If you see your teacher hesitate, put away the mobile and reach for the stopwatch.

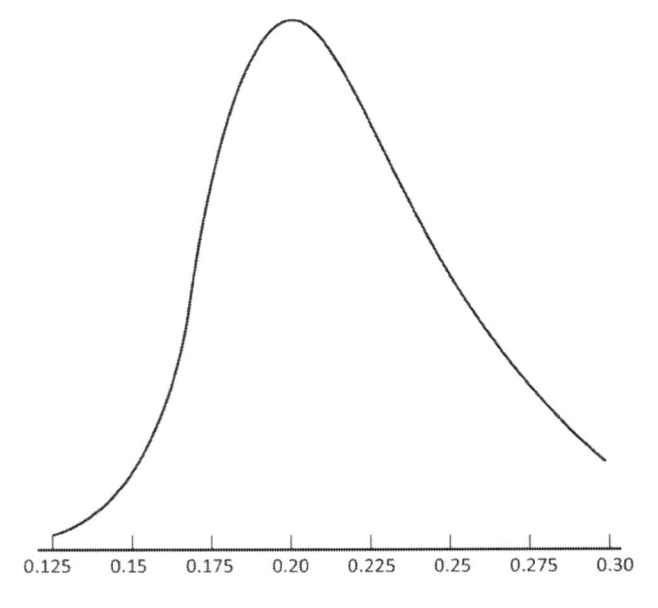

The problem with human timing, of course, is reaction time. This varies from person to person, what mood you're in, what time of day it is, how well you can see the thing you're timing, etc. The graph on the left was assembled from a dataset of over three hundred thousand people (the scale along the bottom is in seconds). It shows the reaction time between being presented with a stimulus and acting upon it. The most common reaction time is 0.20s and the average is 0.212s (the difference being due to the slight skew towards the right).

Reaction time experiments have also revealed what time of day you're at your sharpest (10 o'clock in the morning) and at your doziest (8 o'clock in the evening). Most people are quicker (by about 3ms) on a Friday compared with a Sunday.

Removing the human element from the measurement is obviously much better, though removing the human entirely may not be desirable. When an event starts, it isn't always a crisp, clear moment in time. Take the ubiquitous light gate and timer shown below.

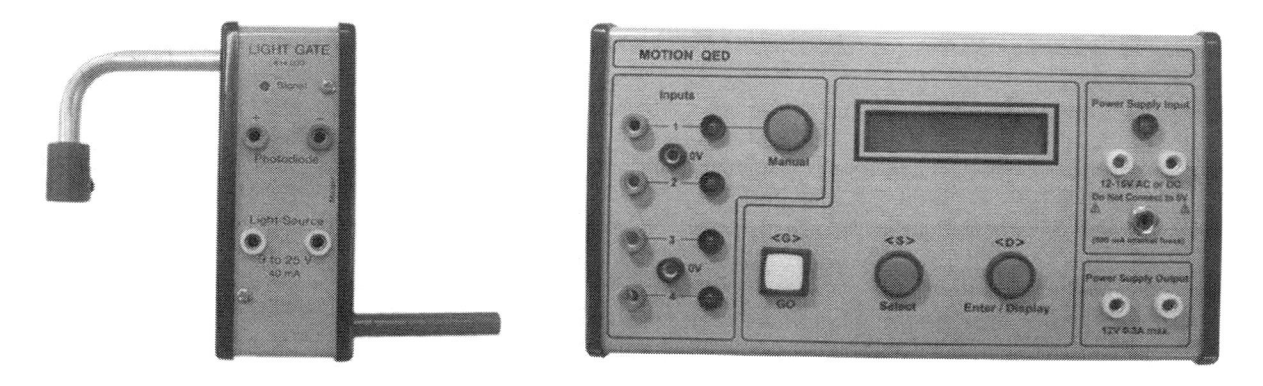

The beam from the light gate has a certain width where the card passes through. This gradually decreases the signal received at the photodiode; rather than a straight 'wham, it's off in an instant' type of response. The electronic timer (the Motion QED box in this example) is monitoring the voltage across the two input leads from the light gate, and will be designed to trigger the timer at a certain voltage (usually 1 volt). Where is the edge of the card in relation to the light beam at this moment? For most experiments it won't matter since the beam is narrow and the card is usually quite wide. But suppose it isn't a wide card; what if it's a 3mm diameter metal rod that's cutting the beam?

The diagram below shows the light level at a 1mm^2 square photodiode as a 3mm diameter rod passes through the light beam from a 3mm wide source (assumed uniform intensity across its width). The rod passes midway between the light source and photodiode, and the distance scale shows the movement of the rod from left to right (with the zero where the rod is centred and no light gets to the photodiode). If the timer triggers on a high light level, the effective diameter of the rod is greater than its actual 3mm (it would be 4mm in the example on the diagram). If the timer triggers at a low light level, the diameter is less (about 2mm in the example).

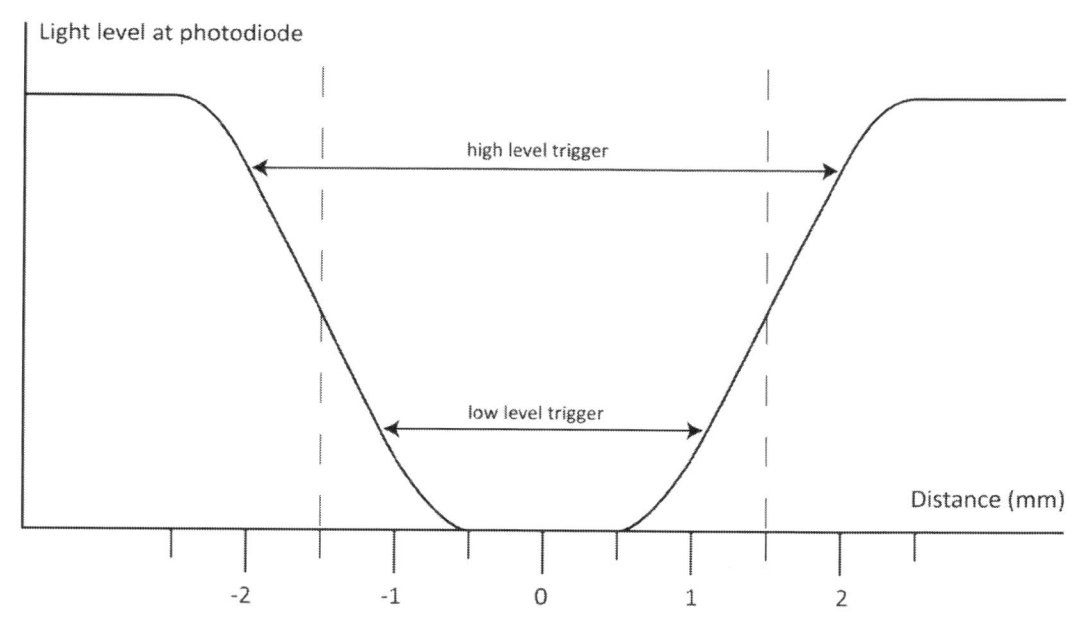

If you are investigating a regular motion, then frequency is another aspect of timing (remember $f = 1/T$). The frequency is the number of cycles of the motion each second and if the frequency is low enough (say, about 1Hz or less), you could hand time it over a large number of cycles. This would be satisfactory for things like a freely swinging pendulum, but it won't work for, say, a vibrating hacksaw blade.

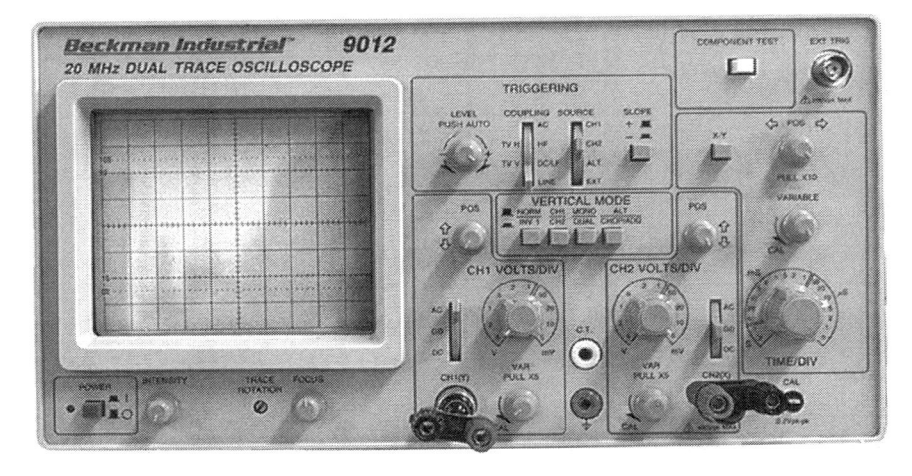

Electrical oscillations in a circuit are easy to monitor; use a suitable multimeter (like the one above left), or simply connect to an oscilloscope. The 'scope above can display two waveforms on the screen at the one time. This is why it's called a dual trace oscilloscope (although it's a bit of a trick since it rapidly chops a single beam and sends half to one waveform and half to the other; an oscilloscope with two electron beams behind the screen is called a dual beam oscilloscope). Both traces on the screen are controlled by a single timebase (so you can't have the spots going across the screen at different speeds), but each one has its own y-gain amplifier, denoted by 'CH1' and 'CH2'. Have a close look just above the timebase and you'll see a control with the word 'cal' (for calibrate) beside it. For frequency measurements using the x-grid on the screen, you must rotate the control to the 'cal' position (quite often it clicks into position).

The frequency of some mechanical oscillations can be monitored by a light gate system connected to a suitable multimeter like the one above. If you can attach a small magnet to the object oscillating, a coil (connected to an oscilloscope) beside the magnet, will give you a picture of the motion as well as the frequency.

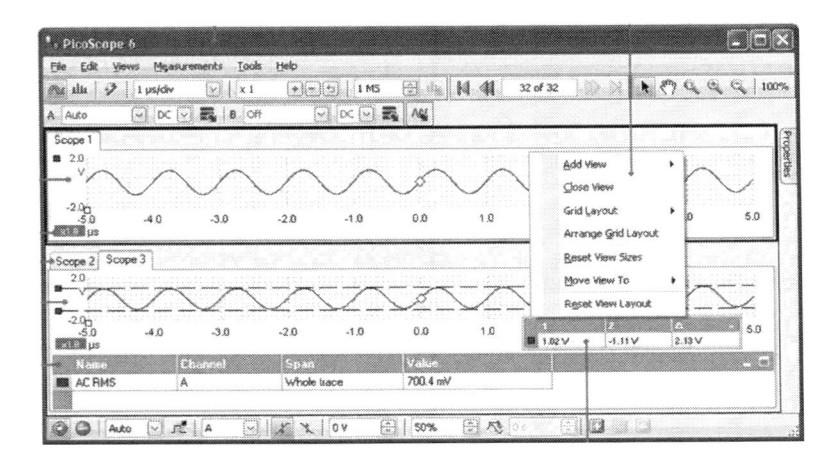

A more recent development is the 'PC oscilloscope'. This isn't one specially designed for liberals; it's just a small box of electronics, a cable for connection to your laptop, and some software for your hard drive. It can capture and analyse short sections of any waveform, and do a frequency analysis of complicated patterns like musical sounds!

Temperature Measurements

In a school setting, this will consist mainly of three instruments: the liquid-in-glass thermometer, the thermistor and the thermocouple. The liquid-in-glass (LiG) thermometer seems straightforward enough; just dip it into the liquid and read the scale. And this is what most people do. But for precise measurements, it's not good enough.

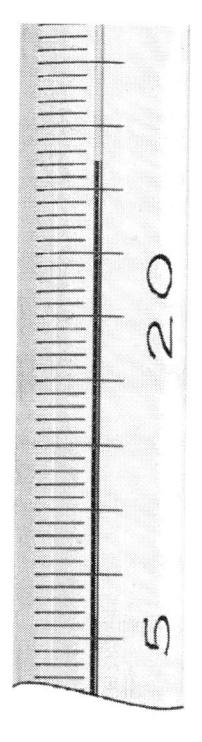

The best thermometers used mercury as the liquid (like the one on the left) but are now rarely seen in schools since mercury was declared a health hazard (many teachers will remember rolling mercury beads across a bench when they were at school; but no, that doesn't explain anything!). Used in a clean capillary tube, its non-wetting properties were much appreciated (non-wetting means that on its way back down the capillary, none of the mercury 'wets' the inside and gets detached).

All the ones used by students will contain alcohol or toluene with a few percent aniline dye added to give the red colour. The two photographs on the right show the top and bottom of such a thermometer beside a ruler. Notice that the scale at the higher temperature is more widely spaced to accommodate the non-linear thermal expansion of the liquid.

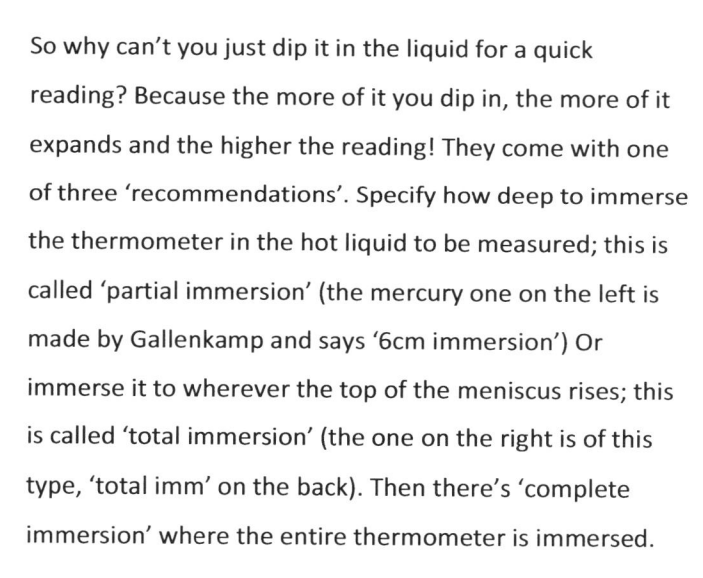

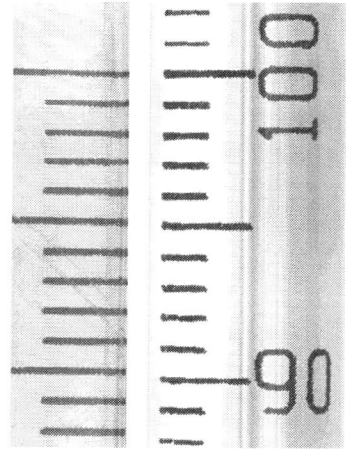

So why can't you just dip it in the liquid for a quick reading? Because the more of it you dip in, the more of it expands and the higher the reading! They come with one of three 'recommendations'. Specify how deep to immerse the thermometer in the hot liquid to be measured; this is called 'partial immersion' (the mercury one on the left is made by Gallenkamp and says '6cm immersion') Or immerse it to wherever the top of the meniscus rises; this is called 'total immersion' (the one on the right is of this type, 'total imm' on the back). Then there's 'complete immersion' where the entire thermometer is immersed.

Immersing only the reservoir of a toluene filled thermometer into liquid at 90°C, which should be used with the 'total immersion' method, could underestimate the true reading by up to 3°C. The correction to be made for this situation is: $\Delta T = 0.001 N (T_{Re} - T_{AV})$, where '$N$' is the number of degrees between the top of the hot liquid and the meniscus, and the bracket is the difference in temperature between the reading and the average temperature of the capillary above the hot liquid.

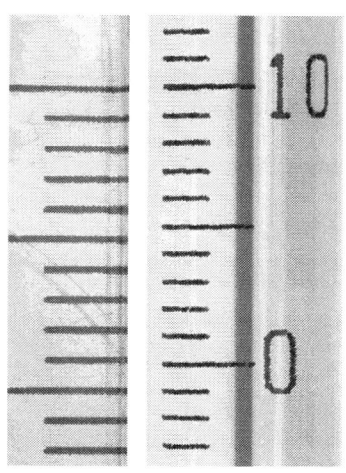

The resistance of an electrical conductor will change if its temperature changes, providing the basis for another type of thermometer. It could be a metal (platinum is common), giving what's called a 'resistance thermometer'; or a man-made ceramic (giving a '**thermistor**'). The former is often used for higher temperatures; the latter being most convenient for room temperature measurements. The thermistors in schools are usually of the type which decreases in resistance as the temperature increases (though you can get ones which do the opposite). They are commonly formed in a disc shape (see page 40 photograph) and can be used up to about 150°C. The potential problem with all of these, is that you have to pass a small current through them to measure the resistance, thereby adding more heat energy to the surroundings (often this is small and can be neglected). A typical response from a thermistor is given in the graph on the right. The ones you use will have the same shape with a different resistance scale. Notice how sensitive it is to a change in temperature; this is its greatest advantage.

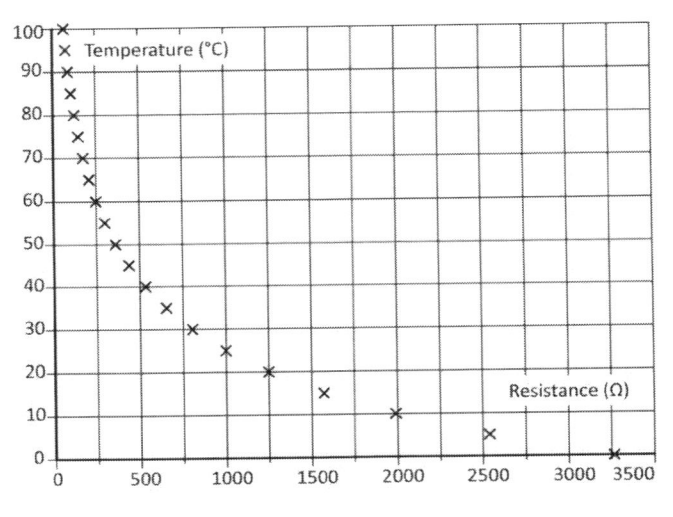

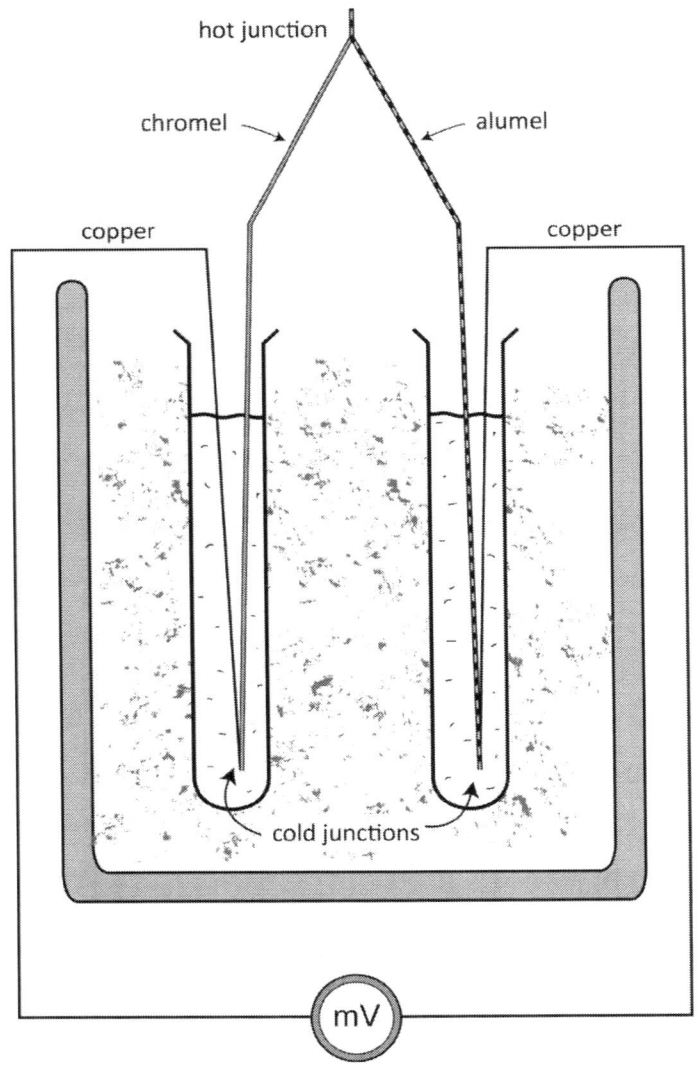

Thermocouples depend on a different effect; the generation of an emf when two dissimilar metals are in a circuit with a temperature gradient. From Standard Grade, you'll remember holding two twisted wires in a bunsen flame, with a multimeter showing the output. This is okay for showing that the blue bit is hotter than the yellow bit of the flame, but no use for measurement since a thermocouple shows the *difference* in temperature between the hot and cold junctions; what you should do is hold the cold function at a fixed, known temperature (usually melting ice). The diagram on the left shows the correct arrangement. Copper and constantan are often used in schools for the two wires, but the best combination uses chromel and alumel. It'll generate over 40μV for every °C of temperature difference. Thermocouples are less sensitive than thermistors, but operate over a much wider range of temperatures.

Light Measurements

The two common detectors used in schools to measure 'light level' are the photodiode and the light dependent resistor (LDR). I've put the words 'light level' in quotes since the relation between your meter reading (in µA or Ω) and a quantitative description of the light absorbed by the detector, is more complicated than you might think. See later for a short discussion of this minefield.

The **photodiode** shown on the top right is an IPL 10020 and is very common. If you acquired it from RS you'll see the letters RS 305-462 around the side. The square in the centre of it is 1mm on a side with about ⅔ of it being light sensitive. It's very useful for its fast response times; it'll react electrically to a sudden change in light level as brief as 10ns (slightly faster on the rise, slightly slower on the fall). Used in reverse bias mode (connect the 'triangle' side of the symbol to the negative side of a dc supply of a few volts), it'll pass currents of the order of microamps at a typical classroom light level.

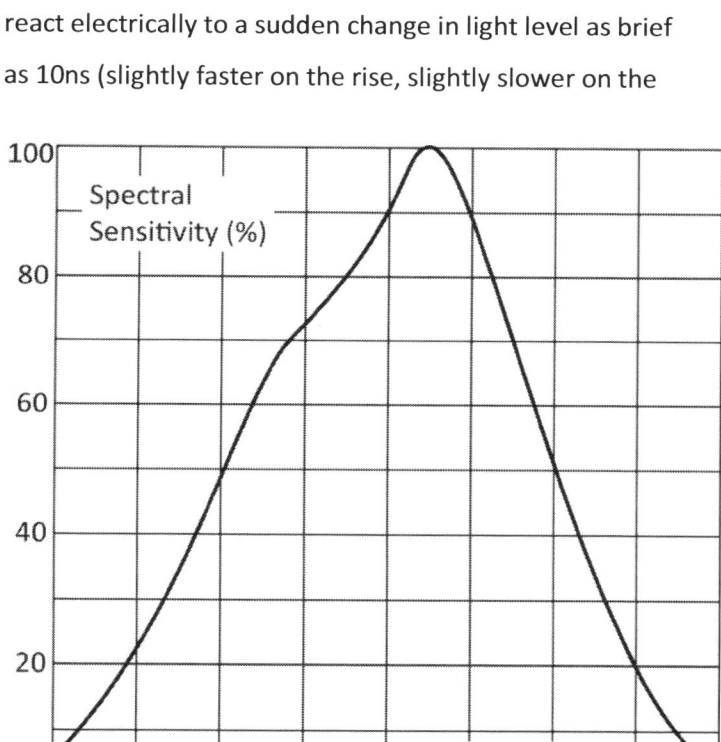

Photodiodes have the very useful property of having a linear response; they pass an output current which is directly proportional to the 'light level' (unlike the LDR).

The graph shows how sensitive it is to light of different wavelengths. You can see that it's not very good for detecting the blue end of the visible spectrum but better at the red end.

In an **LDR**, the 'R' stands for 'resistor', and like all resistors the flow of current doesn't depend upon 'how easily electrons can pass through a material' (an unfortunate impression acquired by pupils, conjuring up images of electrons dodging atoms). In a solid, an electron can only go somewhere if there's a vacant 'energy slot' for it to move into. LDR's have vacant slots, and shining light on the LDR moves electrons into them, producing what we describe as 'conduction'. The most common LDR found in schools is called an ORP-12 (now NORP-12). It costs under £2 if bought loose or £6 if bought mounted. The photograph on the next page shows that the top surface is covered in a squiggly line. It's the light sensitive material - cadmium sulphide (CdS). You can see the ends of the two wires (sticking out like little islands) which connect the LDR to the outside world. The light falling on the CdS squiggle (and around the circumference) controls the resistance between these two points. Attach the wires to a battery, shine light on the LDR, and a light-dependent current is permitted to flow.

The shape of the sensitivity curve is not unlike that of the photodiode with its sharp peak. The rapid drop-off at the blue end occurs when the photon energy (about 2.4eV) promotes valence electrons into the conduction band and fills all the 'energy slots' (thus cutting off conduction). The peak sensitivity is in the yellow region as opposed to the infrared of the photodiode. This makes it possible to mimic the response of the human eye to different wavelengths (making it ideally suited to photography). The LDR is very sensitive to small changes in 'light level', but unfortunately in a non-linear way (the log-log graph though is fairly linear – see graph below right). The NORP-12 comes presented in a hard plastic package; partly to protect the sensitive layer, and partly because all cadmium compounds tend to be dangerous to humans.

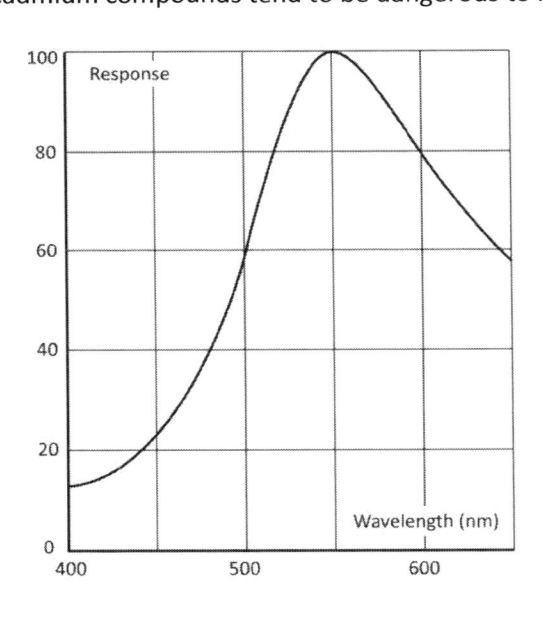

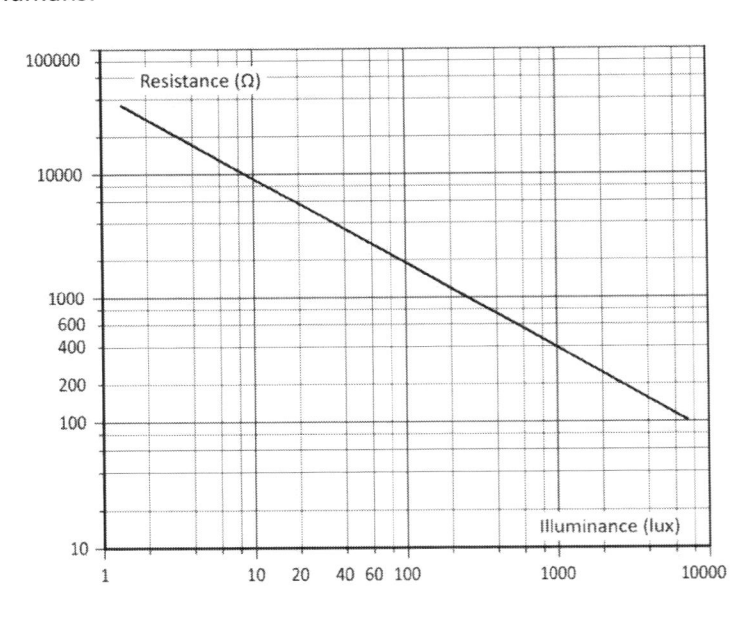

Finally, a word on the minefield of units used to quantify light measurement. There's the scientific side with watts, seconds and metres-squared (called radiometry); but because light is important to humans, there's a parallel set applicable to human perception (called photometry). For example, a distant star in the night sky has the same surface 'brightness' as our own Sun, but you don't need to rub on lots of 'starcream' to avoid 'starburn'. It doesn't help when many of the words are very similar, like luminance and illuminance (measured in 'nits'!!! and lux); or that you'll still come across old units like 'foot-lamberts'. The good news is that you find out about the steradian.

Digital Photography

Over the years, many students have used a camera to take photographs of their experiment laid out on the laboratory bench. With an old chemical film-roll camera, it would have been processed and printed at the chemists shop then glued onto the A4 paper of the Report. With a newer digital camera, it can be downloaded to a computer then captioned and pasted into the Word document, or printed straight from the camera and glued on. If that's all that's required, then many mobile phones have a camera with a good enough resolution to offer an inexpensive alternative. However, the introduction of digital cameras has opened up another possibility; its use as a scientific instrument where you can make measurements.

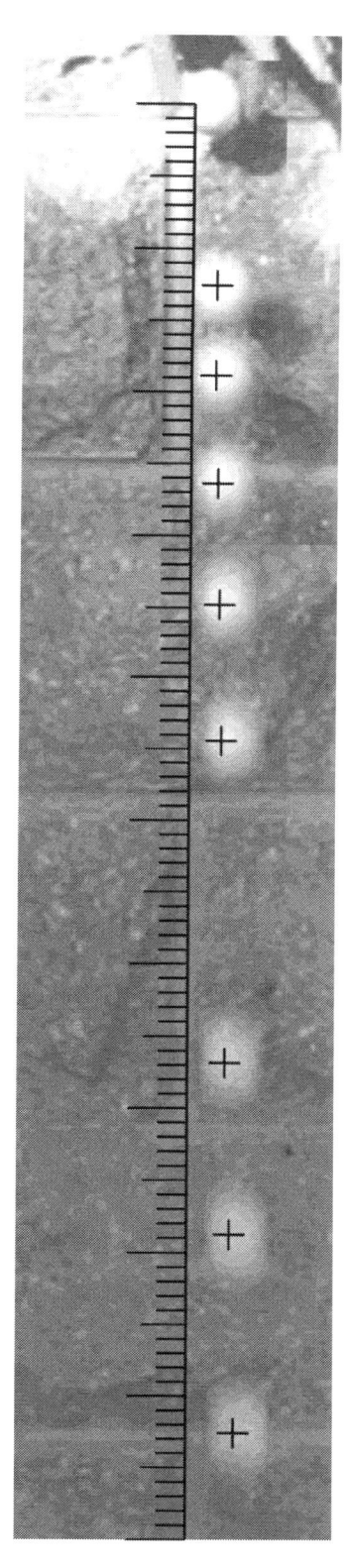

Some digital cameras will have a zoom lens, often allowing close-up pictures to be recorded. To be of any real value as an instrument, it needs to be mounted on a tripod to keep it steady, or at the very least, held against something solid. Even better is to use the shutter release delay function with a tripod, ensuring that pressing the shutter release doesn't wobble the camera. Use it to measure things like the contact angle of a water drop in a surface tension experiment or that difficult-to-read vernier scale. Photograph the waveform of a musical instrument on an oscilloscope screen, or the parabolic curve of an electron beam in an e/m tube. There are many uses; what they all have in common is that you can analyse the results at your leisure. You don't have to hold your neck at a funny angle or screw-up your eyes and get a sore head; you can take your time and make a measurement in a relaxed environment. Examine the results in detail on the computer screen (assuming you took a sharp picture in the first place – spend a bit of time getting it focussed properly). If you intend making a distance measurement from the photograph, take the photograph with a ruler along the bottom of the frame. This will reduce the effect of distortion from the lens.

Many digital cameras have a video facility. You can use it to assemble a time lapse sequence of a dropped ball. You won't need 4k resolution at 60 frames per second for run of the mill applications (but if you've got it, make use of it for its precision). The photograph on the left shows a golf ball being dropped against a block wall and is the result of stitching together the individual frames from the movie. The resolution was only 640 x 480, captured at a setting of 30 frames per second. Most cameras with video facilities will be able to do much better. Look closely and you'll notice a gap about half way down where there should be an image of the golf ball. Your camera will have a memory buffer where it dumps data as it is recorded. It has a size limit. Shooting jpegs won't fill the buffer anytime soon, but shooting in RAW format will typically fill up after 20 to 30 images. The gap mentioned above was when the camera read and flushed the buffer memory. Bear this in mind and read the manual for buffer sizes.

To calculate the acceleration due to gravity, you have to make measurements of time and distance. There is a slight complication in the analysis since the golf ball isn't at x = 0 when t = 0. This means you can't plot distance against time squared (for a slope of ½g). Here's how to handle it:

$$s = \frac{1}{2}g(t+\Delta t)^2 \quad \Rightarrow \quad \sqrt{\frac{2s}{g}} = (t+\Delta t) \quad \Rightarrow \quad t = \sqrt{\frac{2}{g}}\, s^{\frac{1}{2}} - \Delta t$$

A plot of time up the y-axis and the square root of the distance along the x-axis will give a gradient of $\sqrt{2/g}$ with an intercept below the origin of Δt (the time between the ball being dropped and the first image). Writing the equation this way means that the first image with a cross is taken as time zero, but with no adjustment to the distance (no Δs in the equation) you must still align the metre scale with the initial drop position.

The distance data comes from superimposing a metre scale on the photograph and marking the centres of each ball-image with a cross. The time data requires a check on the frame rate of the camera. Taking a movie of a digital timer with an LCD screen gives a blur on the two least significant digits. A better alternative is to use an old fashioned analogue timer; the sort where the hand rotates once per second.

Stop watches of the type once common in athletics are a possibility, but many departments will still have a device called a Venner timer (originally developed at Glenalmond in Perthshire). They are accurate. Test it against your wristwatch to be sure. Here are two stills taken from a movie sequence showing the Venner timer at the start and 30 frames later.

The right hand pointer is almost back to its start position; the actual time between the two photographs being about 0.995s. This gives a time between frames of 0.0332s. The blurry spread of the hand gives a measure of the exposure time of the frame. In the photograph of the dropped ball, this makes the golf ball fuzzy and elongated as it gets faster; limiting the distance over which you can measure to about one metre (if you dropped it from a typical ceiling height, the image blur would be over 20cm long near the floor). The results are plotted on the graph below. The points fit the best-fit straight line quite well, confirming our suspicions that the camera missed a frame (during which it writes the buffer data to the SD memory card).

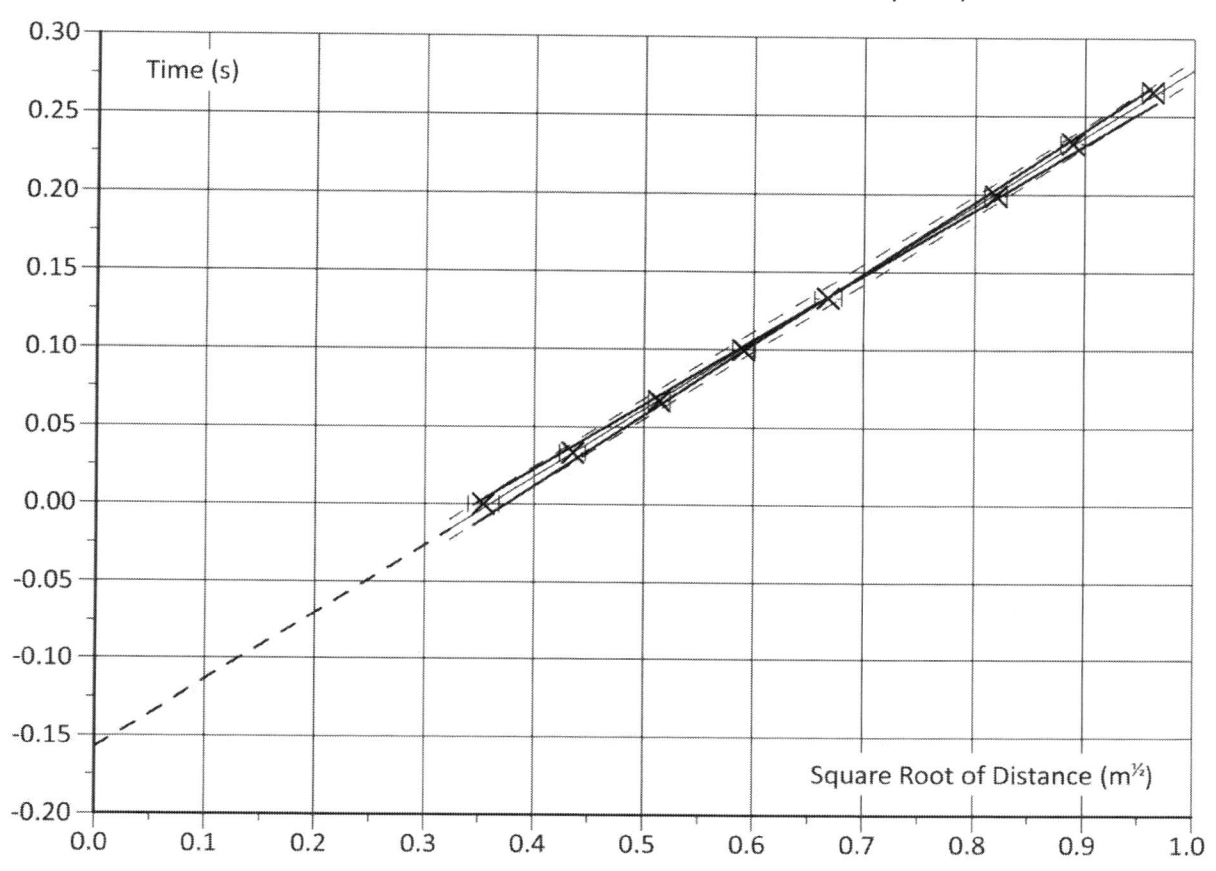

The dotted line shows the projection back to the intercept, showing that the ball was released about 0.16s before the first frame. Using the computer software, the best-fit line gave a slope of 0.4374 ± 0.0033. This corresponds to an acceleration of 10.46 ± 0.16 m s^{-2}. It doesn't take account of any error bars in the readings. The graph shows error bars on the distance scale corresponding to the blurred image of the golf ball (the error bars for the time are too small to show). The simplest analysis using the diagonals of a parallelogram is shown and gives maximum and minimum values of the acceleration of 11.5 m s^{-2} and 9.53 m s^{-2}, for a final result of 10.5 ± 1 m s^{-2}. Factoring in the number of data points would reduce this uncertainty.

Digital SLR cameras also have a rapid fire facility. It's known as 'machine gunning' or 'spray and pray', and makes the whirring noise when the paparazzi photograph a celebrity. It solves the problem of the fuzzy image since you can set the exposure time to 1/500ths and photograph the ball over a longer distance (after a 5m drop the centre of mass of the 4cm diameter ball would move only 2cm). One thing to consider is the calibration of the frame rate stated in the manual. Is the rated 8fps accurate?

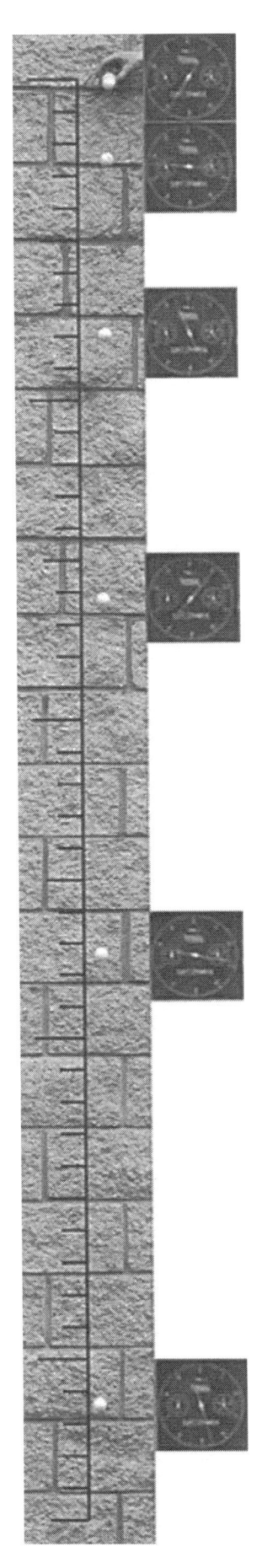

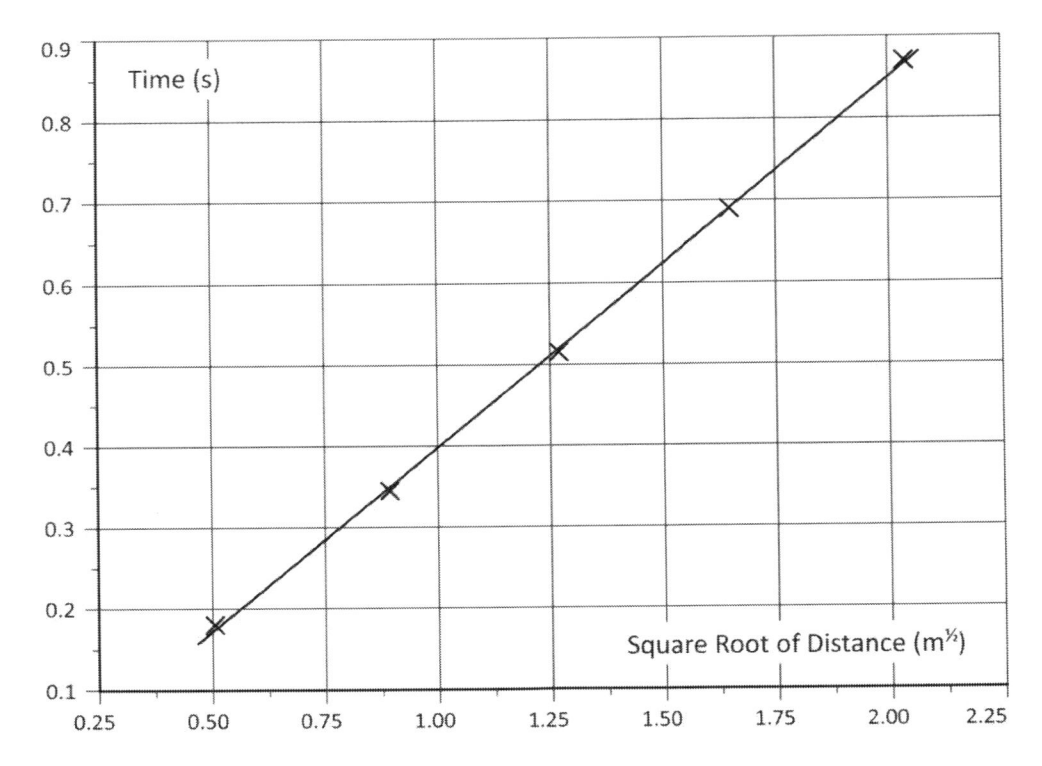

The photograph on the left is a composite of 7 frames shot at 1/500ths. The camera recorded the frames at about 6 per second. It simply gives you more data points for the graph. You'll find that the frame rate is quite variable under different conditions, so it's essential to show a timer in the frame (don't assume it's what's given in the camera spec). I used the same Venner timer as before. With a 5m drop like the one opposite you have to stand well back from the wall with a standard lens to show the full distance. Placed beside the wall, the Venner timer is quite small, so you must take extra care over focussing (you could try placing it closer to the camera and using a small aperture like f22 to get the depth of field, but this will need good lighting conditions to maintain the short exposure).

Using the same variables as before (for the same reasons) gave the graph at the top of the page. Unlike the movie sequence, there are no gaps since the camera can buffer all the frames without reading the data to the CF card. The computer best-fit line gave a slope of 0.45168 which corresponds to an acceleration of 9.803 m s^{-2}. The error bars on the square root distance data are ±0.01 at worst (for the last point), and a constant ±0.01 on the time data, corresponding to an uncertainty of about ±0.5 m s^{-2}.

Try dropping two different sizes of ball and calculating the difference in their accelerations.

Handling Uncertainty 6

Although we will be exploring mathematical and statistical procedures that are used to determine the uncertainty in an experimentally measured quantity, you will see that these are often just 'rules of thumb', and sometimes a good experimenter uses his or her intuition and common sense to simply guess.

………. David Harrison, Department of Physics, University of Toronto

Some experiments try to establish the relationship between two variables, as in Newton's 2nd Law. Others try to measure a physical quantity, like the specific heat capacity of water. The former is often shown on a graph and by an equation. The latter requires your best estimate for the value of the physical quantity, and *an indication of how accurate it is*. This is where uncertainty comes in. To do an experiment, you:

- Decide which method to use
- Use a particular set of instruments
- Read the instruments
- Repeat the experiment several times

The first one involves judgement, but no numbers (and is covered in the marks scheme under 'Procedures'). The other three (in order) cover calibration, reading and random uncertainties. These are the ones you have to quantify. I'll briefly discuss each one in turn, then describe how to combine them into a single final answer.

Calibration Uncertainty

This is a measure of how accurate the instrument is. In theory, you just look up the datasheet which came with the apparatus, or failing that, the Internet site of the manufacturer. In practice, it doesn't always work out like that; the apparatus might be so old that the manufacturer has gone out of business, or it's so downmarket that the manufacturer hasn't bothered to calibrate it. Don't turn your nose up at old equipment though; it's often designed with care and well made. Here's an example of a good quality, ancient, decade resistance box.

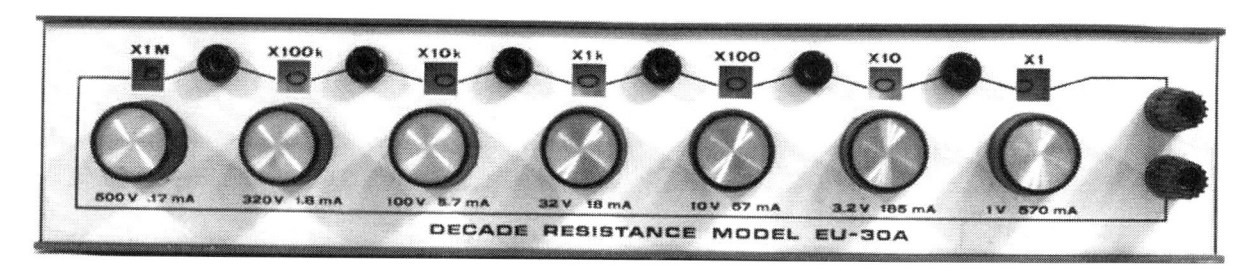

It's got 1960's written all over it, and you'll be lucky if any calibration datasheet is still available since the physics department might have had six principal teachers in the meantime (and if there's one thing new principal teachers like to do, it's have a good tidy-out). So what do you do?

In the case of the resistance box, you could open up the back and check the colour of the tolerance band on the resistors (1%, 0.1% etc), but usually it's not that easy. A common problem is the typical signal generator. They have a frequency scale (analog or digital) which leaves a lot to be desired. The solution is to measure its output frequency with an accurate digital multimeter and take the uncertainty as that of the multimeter (usually always available on a website, and typically giving accuracy like 14.2 ± 0.1 Hz). The dials on these school signal generators are often 'a bit insensitive', and it's tricky setting the frequency to what you want.

Lacking any other information and having done a fruitless web search for datasheets, you just have to state in the Report that the calibration uncertainty is unknown. Explain the problem and justify whether or not you can ignore the uncertainty or otherwise. You can hardly dismiss the calibration uncertainty of what's obviously a piece of cheap junk. If you've made an honest attempt at it, the SQA marker won't penalise you.

The National Qualifications Curriculum Support group produced a booklet in 2001 entitled 'Uncertainties' (published by Learning and Teaching Scotland, ISBN 1859558704) where they suggest values for the calibration uncertainty of common instruments (the whole document repays careful study):

Instrument	Calibration Uncertainty
Wooden metre stick	±½mm
Steel ruler	±0.1mm
Vernier callipers	±0.01mm
Micrometer	±0.002mm
Standard masses (balance)	±5mg
Mercury Thermometer	±½°C
Analogue voltmeter/ammeter	±2% of full scale deflection
Digital meter (volt/amm)	±0.5% of reading + 1 digit
Digital Quartz timer	typically ≈ 1s over 24hrs
Decade resistance box	±1% or 0.1% of value
Signal Generator	±5% of full scale frequency

Reading Uncertainty

Digital Instruments

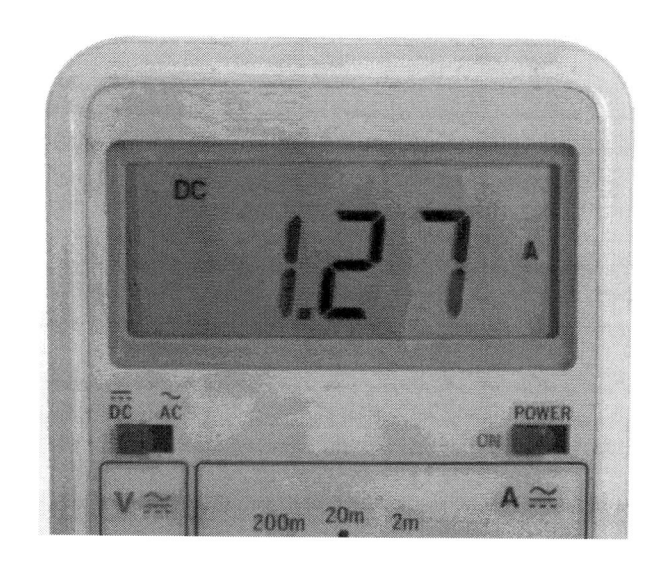

Let's say your digital ammeter is displaying the reading, and you just have to look at it and note it down. What could be simpler? The seven segment displays on the digital ammeter are right in front of you shouting out a reading of 1.27A. What's the problem? It's not 1.28A and it's not 1.26A, it's 1.27A. Here's the problem: the actual current in the circuit could be anywhere between 1.265A (in which case the digital meter would round up the answer) and 1.274999... (where it would round it down). For all values of the actual current in the circuit between these limits, the digital meter will show 1.27A. So the reading uncertainty is ±½ the last digit.

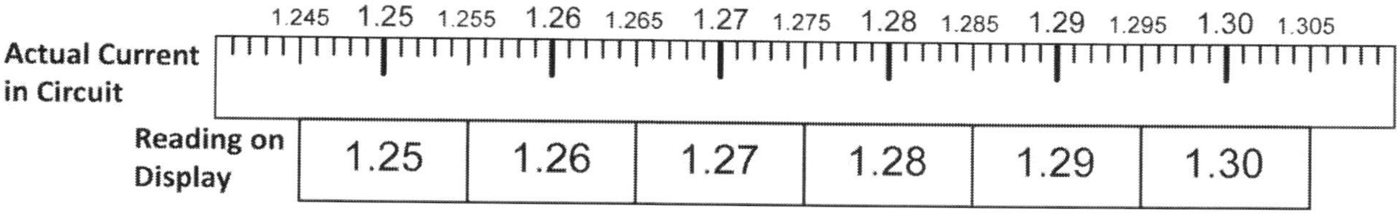

Analog Instruments

An analog instrument is one that isn't digital. Voltmeters with a pointer, a metrestick and a grandfather clock are all examples of analog instruments. The uncertainty from reading these devices depends upon the person taking the reading. The reading uncertainty for your experiment isn't what it tells you in a book; it's what it is for **you** doing **your** experiment. The metrestick for example, is often quoted as ±½mm, but this is just most people most of the time. If you can't get your nose up against the scale, then ±1mm might be appropriate.

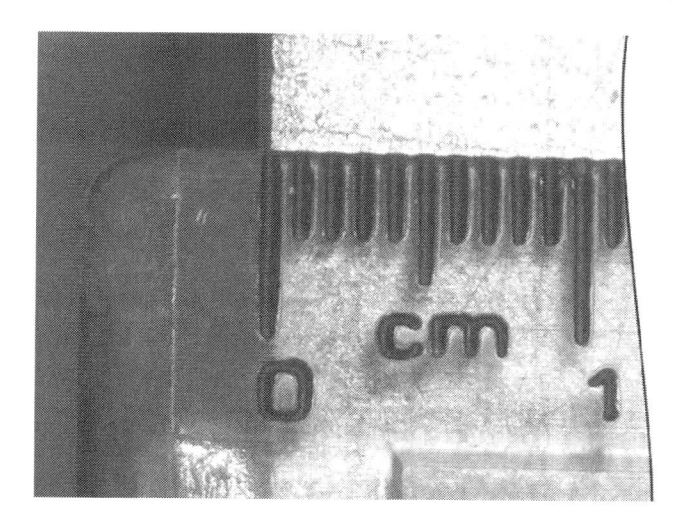

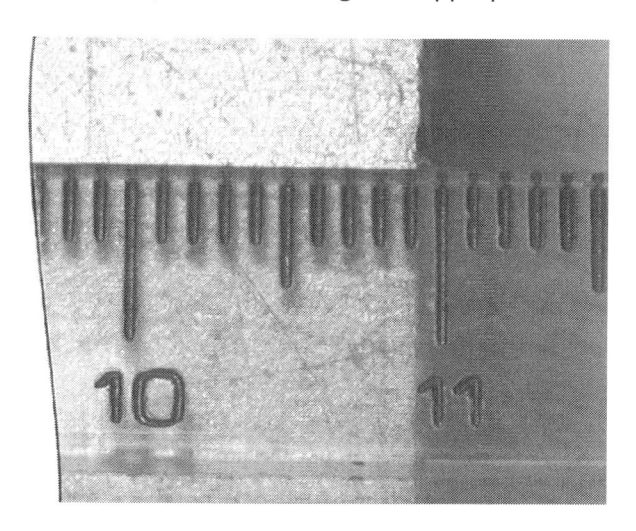

The close-up photos above, remind you that you should take care on lining up the zero of the scale (arguably just as important as the reading at the other end), and that the thing you're measuring may not have a nice clean edge (like the soft aluminium metal used in the photos). What would be your reading for the one on the right: 10.9cm, 10.9cm and a bit? What uncertainty: ±½cm, ±¼cm? Remember that this is a close up so it's easy to read! Perhaps a 5cm convex lens used as a magnifying glass, would come in handy for your own experiments? The photo below is the same piece of bevelled aluminium, this time measured with a different ruler. The zero is aligned with the outside of the bevel and this time the width looks very close to 11cm. Notice how the millimetre marks are painted on the underside of this ruler whereas the other one had its marks grooved on the top. Which one is best?

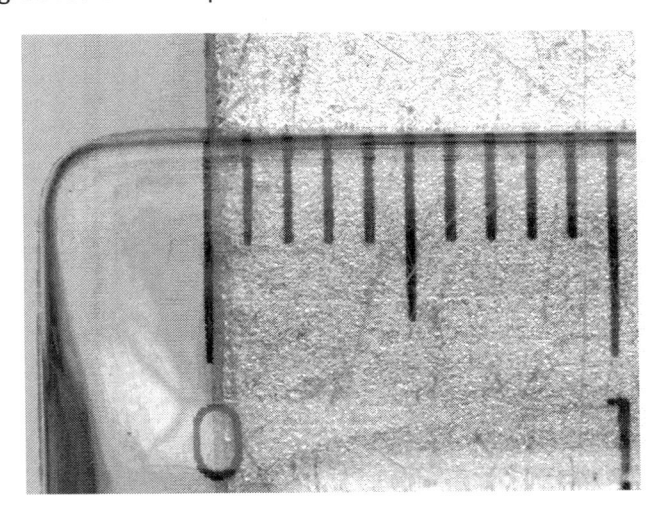

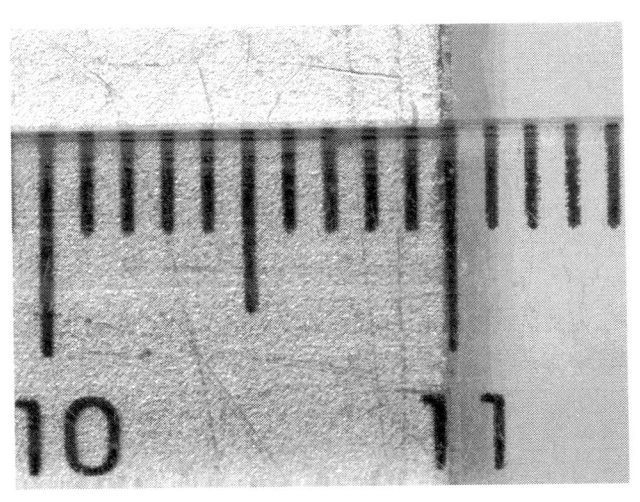

Many schools still have analog voltmeters and ammeters. The tip of the pointer is somewhere on the scale and you've to take a reading. Since the tip of the pointer sits above the scale, you have to take a reading with your eyes above the pointer, not off to the side. This difference due to the change in viewing position is called 'parallax'. It's the effect you get when you hold up a single finger at arm's length in front of you, then look at it with your left eye only, right eye only. . . . notice it shifts against the background. The Hipparcos satellite used the parallax effect to measure the distances to the nearest stars by recording how much a nearby star moved against the distant star background when the Earth had moved to the other side of its orbit around the Sun.

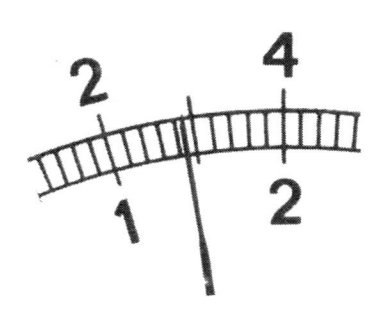

The analog meter below has an add-on called a 'multiplier', used to make the instrument more flexible. The 4mm plug is connected to the input labelled '5V'; which tells you that the maximum reading on the scale is 5volts. So, of the two scales on the display, we read the bottom one. On the close-up, the pointer lies between 1.4volts and 1.5volts. A safe reading would be (1.45 ± 0.05)volts. You might be tempted to read it more accurately, but remember that it depends upon the pointer being zeroed exactly when the meter is disconnected (there's a plastic screw on the facing at the bottom of the pointer for adjustment).

Reading the position of the pointer to half a division either side is what you normally do. With a bit of care, most people could do better; to about ¼ of a division. But, if you think about it, it depends on how wide a division is. If you are using a meter with wide divisions, you might think you could read it to one tenth of a division; but it would be a waste of time. You couldn't locate the pointer accurately in the white space between the division marks; but in any case, the manufacturer is telling you something else. It's a piece of junk. Better instruments have narrower divisions, but this makes it difficult to read to ¼ of a division. The rough rule which covers all analog meters is to take a reading uncertainty of ±½ of the smallest division. A metrestick and a ruler are also examples of analog devices; the thickness of the wood in a metrestick (about 3 to 4mm), places the scale above the piece to be measured (with its accompanying parallax effect).

Random Uncertainty

In some experiments, you can get a more accurate answer by repeating the experiment many times. The idea is that you will probably get as many readings which are too low as are too high, and that your average will be near the correct answer. This doesn't always work though. Ammeters and voltmeters usually show the same readings if you repeat an experiment. But in many cases, especially those involving the human element, repeating the experiment is a good strategy.

Length (m)	Time for 10 swings (s)	Period (s)
0.8	17.93	1.793
0.8	18.06	1.806
0.8	17.81	1.781
0.8	17.67	1.767
0.8	17.82	1.782
0.8	17.98	1.798
0.8	18.01	1.801
0.8	17.72	1.772
0.8	18.07	1.807
0.8	18.02	1.802
0.8	17.73	1.773
0.8	17.62	1.762
0.8	17.80	1.780
0.8	18.22	1.822
0.8	17.94	1.794

How many 'runs' of an experiment should you do? The answer is; as many as you've got time for! An experiment which just needs a simple reset to get it back to the start, could be repeated many times. Other experiments (like viscosity, where there's refilling or fishing), are time consuming to repeat. Doing them twenty times would be a heroic feat, worthy of a special plaque in the Assembly Hall. So there's no set number. However, except in unusual circumstances SQA markers would not consider two 'runs' as sufficient. Three 'runs' is entering the grey area, but five or six keeps you safe from criticism. Ten is common in pendulum experiments. Repeating an experiment is an example of the law of diminishing returns. If you've done the experiment twice, you gain a lot by doing it five more times. If you've done it one hundred times, you gain little by doing it five more times.

Calibration and reading uncertainties don't require any calculations; just a plain statement of the values. If you repeat an experiment many times, you have to use **all** of that data to calculate your best answer with its random uncertainty. The above table shows a classic example. A simple pendulum of length 0.8m is timed over 10 swings (longer time means smaller fractional uncertainty). The experiment is repeated 14 times.

The first thing to do is calculate the mean value of the period:

$$\bar{x} = \frac{\sum x_i}{N} = \frac{26.84}{15} = 1.789 \text{ s}$$

The method you were given in 5th year for calculating the random uncertainty:

$$\Delta x = \frac{(\text{biggest - smallest})}{N} = \frac{(1.822 - 1.762)}{15} = 0.004 \text{ s}$$

gives a reasonable answer up to about a dozen readings (we have 15 readings so it should be fairly reliable).

A better alternative for estimating the uncertainty uses the standard deviation method.

$$\text{std. dev.} = \sqrt{\frac{\sum (x_i - \bar{x})^2}{N}} = \sqrt{\frac{0.00438}{15}} = 0.017 \text{ s}$$

Note: The above formula is sometimes seen with $(N-1)$ instead of N. It's the difference between the whole population and a sample of that population. Use the N version to get the standard deviation of your sample. Using your sample readings with the $(N-1)$ in the bottom line gives an estimate of what the standard deviation would be for the whole population.

The standard deviation gives a measure of the 'fatness' of the histogram of the data. This isn't the same as uncertainty. What we want is the uncertainty in the mean value, which you get by:

$$\Delta x = \frac{\text{std. dev.}}{\sqrt{N}} = \frac{0.017}{\sqrt{15}} = 0.0044 \text{ s}$$

So the answer is: $T = (1.789 \pm 0.0044)$ s. Round the uncertainty to give a final answer of: $T = (1.789 \pm 0.005)$ s.

Notice that there are no error bars in the period column of the table. You would repeat the procedure for different lengths of pendulum, calculating the random uncertainty of the period for each length. You then have to combine the random uncertainty with the calibration and reading uncertainties of your timer to get a total uncertainty in the period for each length. At that stage you could draw-up a summary table showing the length (with its uncertainty) and the period (with its uncertainty). The next section shows how to combine the three types of uncertainty.

Combining Uncertainties

Gather together your values for the calibration uncertainty, reading uncertainty and random uncertainty. For the pendulum experiment, and using the raw data for 10 swings, the uncertainties for one period would be:

Timer: Calibration uncertainty (1s per day over 18s)/10   ±0.00002s

Reading uncertainty (½ of last digit → ½ of 0.01)/10   ±0.0005s

Random uncertainty (from above)   ±0.005s

Stop for a minute and assess them. Do they seem reasonable? The timer used was a quartz type and so the calibration uncertainty is insignificant. The digital display of the timer also gives a small reading uncertainty. The biggest contribution is from the random uncertainty. This is good news since it means you could reduce it further by taking more readings. The way to combine uncertainties is the 'Root Sum of Squares' (RSS) method. For the data above:

$$\text{total uncertainty} = \sqrt{0.00002^2 + 0.0005^2 + 0.005^2} = 0.00502\,s$$

Notice how close the result is to the random uncertainty (which was the biggest). The rough rule is that if one of the uncertainties is more than 3 times any of the others, you just use its uncertainty for the total (though see the discussion on page 121 for the use of this rule with a more complicated expression). The reason for the 3 is that if you square two numbers, and one is three times bigger than the other, it makes the bigger one 9 times bigger. To a physicist, this is the same as 10, which is 'order of magnitude'; and that means you make life easier but still get a good answer, by ignoring the rest.

Here are two examples:

1. Calibration uncertainty   ±0.0005m
   Reading uncertainty   ±0.0005m
   Random uncertainty   ±0.0002m

$$\text{total uncertainty} = \sqrt{0.0005^2 + 0.0005^2 + 0.0002^2} = 0.00073\,s$$

(the SQA would round this down to 0 0.0007s, I would round it up to 0.0008s)

2. Calibration uncertainty   ±0.0001kg
   Reading uncertainty   ±0.0005kg
   Random uncertainty   ±0.0012kg

$$\text{total uncertainty} = \sqrt{0.0005^2 + 0.0012^2} \Rightarrow \pm 0.0013\,kg$$

(SQA down to ±0.001kg, me up to ±0.002kg)

Final Rounding Off

You've spent a lot of effort slaving over the calculator, trying to be precise, trying to get your best answer, when at the last minute we want you to spoil it all by rounding it off. It's one of those times when physics can seem a bit bureaucratic, possibly even daft.

What we're trying to do is extract the maximum information without asking too much of the raw data.

As a rule, don't do any serious rounding until the last step. But notice that we have **two** things to round off; the mean value and the uncertainty. **They are rounded off differently**. The mean value is rounded off in the usual maths way; '5 and above is rounded up'. The uncertainty is only rounded down if it makes a difference of 5% or less, so for example, ±2.1 would be rounded down to ±2 since 0.1 is less than 5% of 2.1, but ±2.2 would be rounded up to ±3 since 0.2 is more than 5% of 2.2. Some people will think this a bit odd, but it is the advice given in the UKAS publication M3003 'The Expression of Uncertainty and Confidence in Measurement', and I agree with them. To satisfy SQA markers who're still thinking along the lines of '5 and above', I would quote this reference in full if you round off the uncertainty this way.

UKAS publication M3003 *The Expression of Uncertainty and Confidence in Measurement* **Edition 1, Dec 1997**

Here are a few examples; most are straightforward but a few show 'grey areas'.

Raw Value	Rounded Value	Comment
4.51±0.18	4.5±0.2	The uncertainty (0.18) is 18 times bigger than the rounding error (going from 4.51 to 4.5 is a change of 0.01) so round off
239±21	240±20	The rounding error (239 to 240) is much less than the uncertainty (21) so round off; it's one of those rare uncertainties that's rounded down
235±10	235±10	Leave the mean value (235) as it is, since the uncertainty (10) is only twice the rounding error (5)
235±53	240±60 or 235±60	More temptation to round to (240) though some would view this as too coarse and leave it at (235)
0.260±0.007	0.26±0.01	The '0' in 0.260 isn't justified since the uncertainty (0.007) is much bigger than the place value (0.001) of that '0'
0.61±0.12	0.6±0.2	The uncertainty (0.12) is much bigger than the least significant digit of the mean value (the '1' in 0.61), so round off / round up
57.4±1.4	57±2 or leave at 57.4±1.4	The uncertainty (1.4) is not much bigger than the rounding error of the mean (down 0.4 to 57). This is one of those grey areas and you'd need more information on the experiment to decide which is best.
0.5±1.5	0.5±1.5	There's nothing wrong with having the uncertainty bigger than the mean. People who follow the rule book would round to 1±2.

The lesson from these examples is to compare the size of the uncertainty with the error in rounding the mean value. Quite often, the uncertainty is a lot bigger than the error from rounding off the mean value. I hope your project gets ones like that. If the uncertainty is about the same size as the error in rounding the mean, then just don't round off the mean value (though you still might need to round off the uncertainty eg. 0.34 ± 0.012 rounded to 0.34 ± 0.02). It's the cases in between that are the difficult ones, where you have to use your judgement. It's a good strategy to add in a sentence explaining your decision since it puts pressure on the SQA marker to accept your choice. At the end of the day, when your fingers are tired and your eyes are sore, most of your uncertainties will end up rounded to one significant figure. You'll need two sig. figs. less often, and three sig. figs. would be a hanging offence.

Calculating the Uncertainty in a Formula

You've got your equation $F = ma$, you measure the mass and acceleration with their uncertainties, and you want to calculate the answer for the force with its uncertainty. This is an example of a product, but we also have expressions with additions $R_T = R_1 + R_2$, quotients $R = V/I$, powers $E_k = \frac{1}{2}mv^2$ and combinations, $s = ut + \frac{1}{2}at^2$. Some of them are handled using absolute uncertainties (like Δx) and some with fractional uncertainties (like $\Delta x/x$). If all the quantities have the same units, you use absolute uncertainties. Using fractional uncertainties is a way of 'levelling the playing field' in what may be quantities of vastly different size. Here are some examples:

1. **Addition/subtraction** $L = l_1 + l_2$ $l_1 = (15.2 \pm 0.1)\,\text{cm}$ $l_2 = (9.4 \pm 0.1)\,\text{cm}$

$$(\Delta L)^2 = (\Delta l_1)^2 + (\Delta l_2)^2$$
$$= 0.1^2 + 0.1^2 = 0.02$$
$$\Delta L = \sqrt{0.02} = 0.14\,\text{cm}$$
$$L = l_1 + l_2 = 15.2 + 9.4 = 24.6\,\text{cm}$$
$$L = (24.6 \pm 0.14)\,\text{cm} \;\rightarrow\; (24.6 \pm 0.2)\,\text{cm}$$

2. **Multiply / divide** $T = I\alpha$ $I = 0.28 \pm 0.03\,\text{kg m}^2$ $\alpha = 0.79 \pm 0.09\,\text{rad s}^{-2}$

$$\left(\frac{\Delta T}{T}\right)^2 = \left(\frac{\Delta I}{I}\right)^2 + \left(\frac{\Delta \alpha}{\alpha}\right)^2$$
$$= \left(\frac{0.03}{0.28}\right)^2 + \left(\frac{0.09}{0.79}\right)^2 = 0.024$$
$$\frac{\Delta T}{T} = \sqrt{0.024} = 0.16 \qquad \Delta T = T \times 0.16$$
$$T = I\alpha = 0.28 \times 0.79 = 0.22\,\text{Nm}$$
$$\Delta T = 0.22 \times 0.16 = 0.035\,\text{Nm}$$
$$T = (0.22 \pm 0.035)\,\text{Nm} \;\rightarrow\; (0.22 \pm 0.04)\,\text{Nm}$$

3. **Powers** $s = \frac{1}{2}at^2$ $a = 0.054 \pm 0.006\,\text{ms}^{-2}$ $t = 2.5 \pm 0.3\,\text{s}$

$$\left(\frac{\Delta s}{s}\right)^2 = \left(\frac{\Delta a}{a}\right)^2 + \left(\frac{\Delta(t^2)}{t^2}\right)^2 = \left(\frac{\Delta a}{a}\right)^2 + \left(2\frac{\Delta t}{t}\right)^2 = \left(\frac{\Delta a}{a}\right)^2 + 4\left(\frac{\Delta t}{t}\right)^2$$

$$\left(\frac{\Delta s}{s}\right)^2 = \left(\frac{0.006}{0.054}\right)^2 + 4\left(\frac{0.3}{2.5}\right)^2 = 0.070$$

$$\frac{\Delta s}{s} = \sqrt{0.07} = 0.26 \;\Rightarrow\; \Delta s = s \times 0.26$$

$$s = \frac{1}{2}at^2 = \frac{1}{2} \times 0.054 \times 2.5^2 = 0.169\,\text{m}$$
$$\Delta s = s \times 0.26 = 0.169 \times 0.26 = 0.045\,\text{m}$$
$$s = (0.169 \pm 0.045)\,\text{m} \;\rightarrow\; (0.17 \pm 0.05)\,\text{m}$$

Three Common Cases:

A. Calculate the period squared (T^2) given the period $T = (1.7 \pm 0.2)\text{s}$.

$$\left(\frac{\Delta(T^2)}{T^2}\right) = \frac{2T\Delta T}{T^2} = 2\frac{\Delta T}{T} \qquad \text{(just like differentiation)}$$

$$= 2 \times \frac{0.2}{1.7} = 0.235$$

$$\Delta(T^2) = 0.235 \times T^2 = 0.235 \times 1.7^2 = 0.235 \times 2.89 = 0.68$$

$$\Rightarrow \quad T^2 = (2.89 \pm 0.68)\text{s}^2 \quad \rightarrow \quad (2.9 \pm 0.7)\text{s}^2$$

B. Calculate $\Delta\lambda$ from $n\lambda = d\sin\theta$ given $n = 1$, $1/d \rightarrow 300$ lines per mm, $\theta = 10½° \pm ½°$

$$10.5° \pm 0.5° \quad \Rightarrow \quad (0.183 \pm 0.009)\text{radians} \quad \text{(no rounding yet)}$$

$$\left(\frac{\Delta\lambda}{\lambda}\right)^2 = \left(\frac{\Delta d}{d}\right)^2 + \left(\frac{\Delta\sin\theta}{\sin\theta}\right)^2 \qquad \Delta d = 0 \quad \Rightarrow \quad \frac{\Delta\lambda}{\lambda} = \frac{\Delta\sin\theta}{\sin\theta}$$

$$\frac{\Delta\lambda}{\lambda} = \frac{\Delta\sin\theta}{\sin\theta} = \cos\theta\frac{\Delta\theta}{\sin\theta} \qquad (\Delta\theta \text{ in radians})$$

$$\frac{\Delta\lambda}{\lambda} = \cos\theta\frac{\Delta\theta}{\sin\theta} = \cos 0.183 \times \frac{0.009}{\sin 0.183} = 0.0486$$

$$\lambda = d\sin\theta = \frac{1}{300000} \times \sin 10.5° = 607\text{nm}$$

$$\Delta\lambda = 0.0486 \times \lambda = 0.0486 \times 607\text{nm} = 29.5\text{nm}$$

$$\lambda = (607 \pm 29.5)\text{nm} \quad \rightarrow \quad (610 \pm 30)\text{nm}$$

Note that it would be just as sensible to assume the wavelength of the monochromatic light and use this method to measure the groove spacing (d) of the diffraction grating.

C. The Moment of Inertia of a solid disc of mass (0.837 ± 0.005)kg and radius (0.115 ± 0.003)m.

$$I = \frac{1}{2}mR^2 = \frac{1}{2} \times 0.837 \times 0.115^2 = 0.0055 \text{kgm}^2$$

$$\frac{\Delta m}{m} = \frac{0.005}{0.837} \Rightarrow 0.6\% \qquad \frac{2\Delta R}{R} = \frac{2 \times 0.003}{0.115} \Rightarrow 5\% \qquad \text{so ignore the mass term}$$

$$\left(\frac{\Delta I}{I}\right)^2 = \left(\frac{\Delta\frac{1}{2}}{\frac{1}{2}}\right)^2 + \left(\frac{\Delta m}{m}\right)^2 + \left(\frac{2\Delta R}{R}\right)^2 = 0 + 0 + \left(\frac{2\Delta R}{R}\right)^2 \quad \Rightarrow \quad \frac{\Delta I}{I} = \frac{2\Delta R}{R} \equiv 5\%$$

5% of 0.0055 is ≈ 0.0003 $\quad \Rightarrow \quad I = (0.0055 \pm 0.0003)\text{kgm}^2$

The General Case

This is a way of handling any formula. It uses a rather daunting looking equation, which is really a bit of an imposter. You have a function with as many variables as you like; here's an example with three, $w = f(x, y, z)$ and you want to determine the uncertainty in the result Δw, given the uncertainties Δx, Δy and Δz. It could contain any combination of multiplying, adding, subtracting, powers; anything you care to throw at it. A simple example from heat energy is $E_h = f(c, m, \Delta T)$, where only multiplication is involved. Here's how to calculate it. Start with the maths expression (ask your maths teacher about it):

$$\Delta w = \frac{\partial f}{\partial x}\Delta x + \frac{\partial f}{\partial y}\Delta y + \frac{\partial f}{\partial z}\Delta z$$

And modify it for use with uncertainties in experiments:

$$(\Delta w)^2 = \left(\frac{\partial f}{\partial x}\Delta x\right)^2 + \left(\frac{\partial f}{\partial y}\Delta y\right)^2 + \left(\frac{\partial f}{\partial z}\Delta z\right)^2$$

This is the same RSS technique which assumes the uncertainties (Δx, Δy and Δz) are not correlated (if they were correlated, use the version without the squares). The funny looking symbols are called partial differentials (again, ask your maths teacher) but we'll treat them just like ordinary differentials, so read $\frac{\partial f}{\partial x}$ as $\frac{df}{dx}$. Here's an example using $s = ut + \frac{1}{2}at^2$ (like $s = f(u,a,t)$). Calculate the differentials:

$$\frac{\partial s}{\partial u} = \frac{\partial}{\partial u}\left(ut + \frac{1}{2}at^2\right) = t \qquad \frac{\partial s}{\partial a} = \frac{\partial}{\partial a}\left(ut + \frac{1}{2}at^2\right) = \frac{1}{2}t^2 \qquad \frac{\partial s}{\partial t} = \frac{\partial}{\partial t}\left(ut + \frac{1}{2}at^2\right) = u + at$$

Let's try it with this data: $\quad u = 28 \pm 3 \text{ms}^{-1} \qquad a = 0.72 \pm 0.08 \text{ms}^{-2} \qquad t = 2.6 \pm 0.2 \text{s}$

$$\frac{\partial s}{\partial u} = t = 2.6 \qquad \frac{\partial s}{\partial a} = \frac{1}{2}t^2 = \frac{1}{2} \times 2.6^2 = 3.38 \qquad \frac{\partial s}{\partial t} = u + at = 28 + 0.72 \times 2.6 = 29.872$$

$$(\Delta s)^2 = \left(\frac{\partial s}{\partial u}\Delta u\right)^2 + \left(\frac{\partial s}{\partial a}\Delta a\right)^2 + \left(\frac{\partial s}{\partial t}\Delta t\right)^2 = (2.6 \times 3)^2 + (3.38 \times 0.08)^2 + (29.872 \times 0.2)^2$$

$$(\Delta s)^2 = 60.84 + 0.073 + 35.69 = 96.6 \quad \Rightarrow \quad \Delta s = 9.8 \text{m}$$

$$s = ut + \frac{1}{2}at^2 = 28 \times 2.6 + \frac{1}{2} \times 0.72 \times 2.6^2 = 75.2 \text{m}$$

$$s = (75.2 \pm 9.8) \text{m} \quad \rightarrow \quad s = (75 \pm 10) \text{m}$$

There's an interesting lesson from this example. The % uncertainties in the raw data were 11%, 11% and 8%. You would expect all of them to be important in the final uncertainty, yet the contribution from the acceleration turns out to be tiny (look at the three numbers that make up the 96.6). It's because the answer comes from two separate parts, ut and $\frac{1}{2}at^2$. With our data, the first part dominates the answer for short times. If the time was much longer, the second term would dominate and the uncertainty in the acceleration would become important. Your old rule of thumb . . . . . *'if one % uncertainty is more than three times any other % uncertainty, then only take'* . . . . . only works for simple cases. If you plan to study for a degree in physics at university, you should get familiar with this general method.

Drawing Graphs 7

Most students will use computer software to construct tables and draw graphs, but you don't have to do this. Doing the tables and graphs by hand is perfectly acceptable; just make sure that, either way, the end result is satisfactory. And this means:

- A4 size except in exceptional circumstances.
- Scale should include the origin unless this is clearly impractical.
- Points are clearly marked (so don't use dots).
- Show 'error bars' (funny how nobody calls them uncertainty bars!)
- Both axes labelled with the name of the each variable in words, and with its units.
- Plenty of gridlines.
- Draw in the best-fit line / curve, either by hand or using the software. You can use the values for the gradient, and the uncertainty in the gradient, given by the software.

Here's an example taken from a pendulum experiment. The graph beside it shows how you can lose marks.

Length 'L' (m)	Period Squared T^2 (s^2)
0.206±0.002	0.83±0.04
0.252±0.002	1.01±0.04
0.303±0.002	1.25±0.05
0.367±0.002	1.48±0.05
0.409±0.002	1.65±0.05
0.455±0.002	1.86±0.06
0.501±0.002	2.02±0.06
0.558±0.002	2.25±0.06
0.602±0.002	2.42±0.06
0.651±0.002	2.58±0.07
0.704±0.002	2.83±0.07
0.757±0.002	3.05±0.07
0.806±0.002	3.26±0.07
0.849±0.002	3.42±0.08
0.905±0.002	3.66±0.08

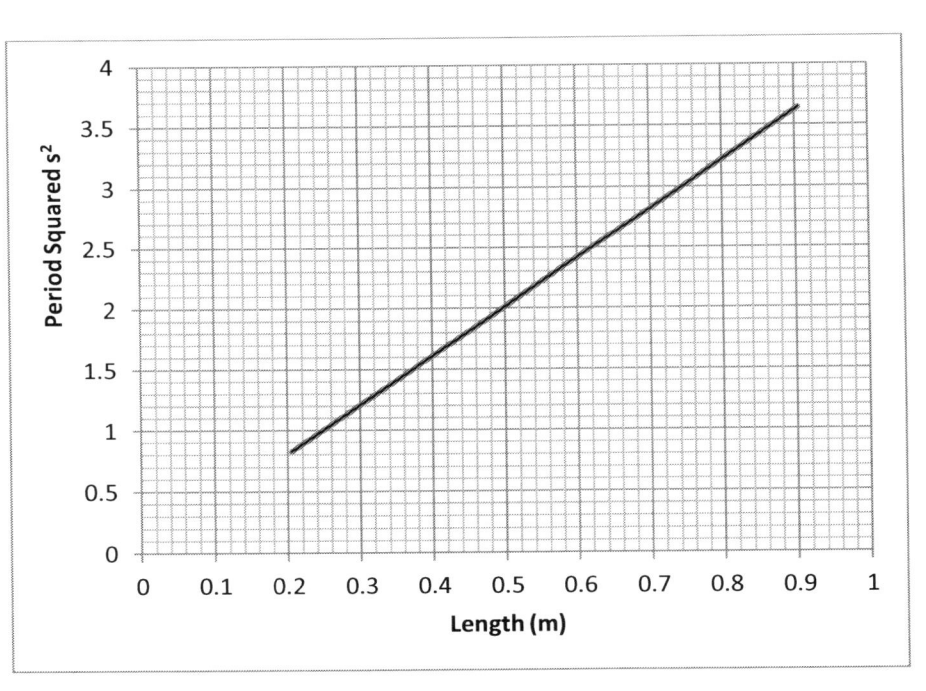

Firstly, it's too small. The size you see in front of you is common in reports. Make the graph A4. Secondly, there are no error bars. For the length data, this is okay since they are too small to show, but they should be drawn for the period squared data. Thirdly, there are no data points indicated. You can't get away with claiming that the computer has drawn the best line, so why do you need to show the data points! Let humans be the best judge, and let them see the points on the graph. The good news is that the gridlines and labels are okay. The next page shows the graph correctly drawn. If it's one of many, add a caption / reference beside it.

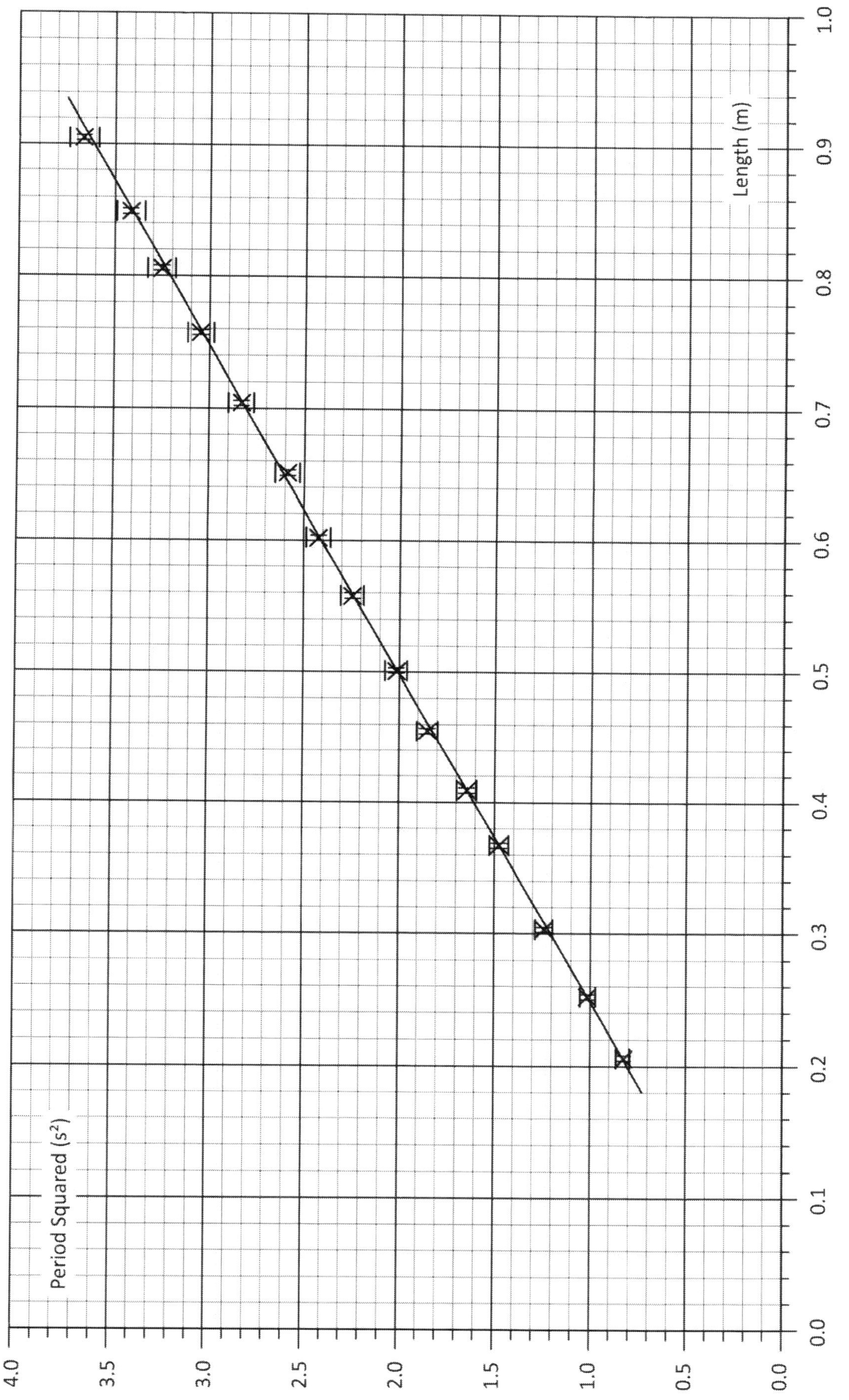
Length (m)
Period Squared (s^2)
0.0
0.1
0.2
0.3
0.4
0.5
0.6
0.7
0.8
0.9
1.0
0.0
0.5
1.0
1.5
2.0
2.5
3.0
3.5
4.0

Here's an example of a poor hand drawn graph. It has no vertical gridlines (it's not even drawn on graph paper) and hasn't got any error bars. The 'amps' should probably be 'milliamps'.

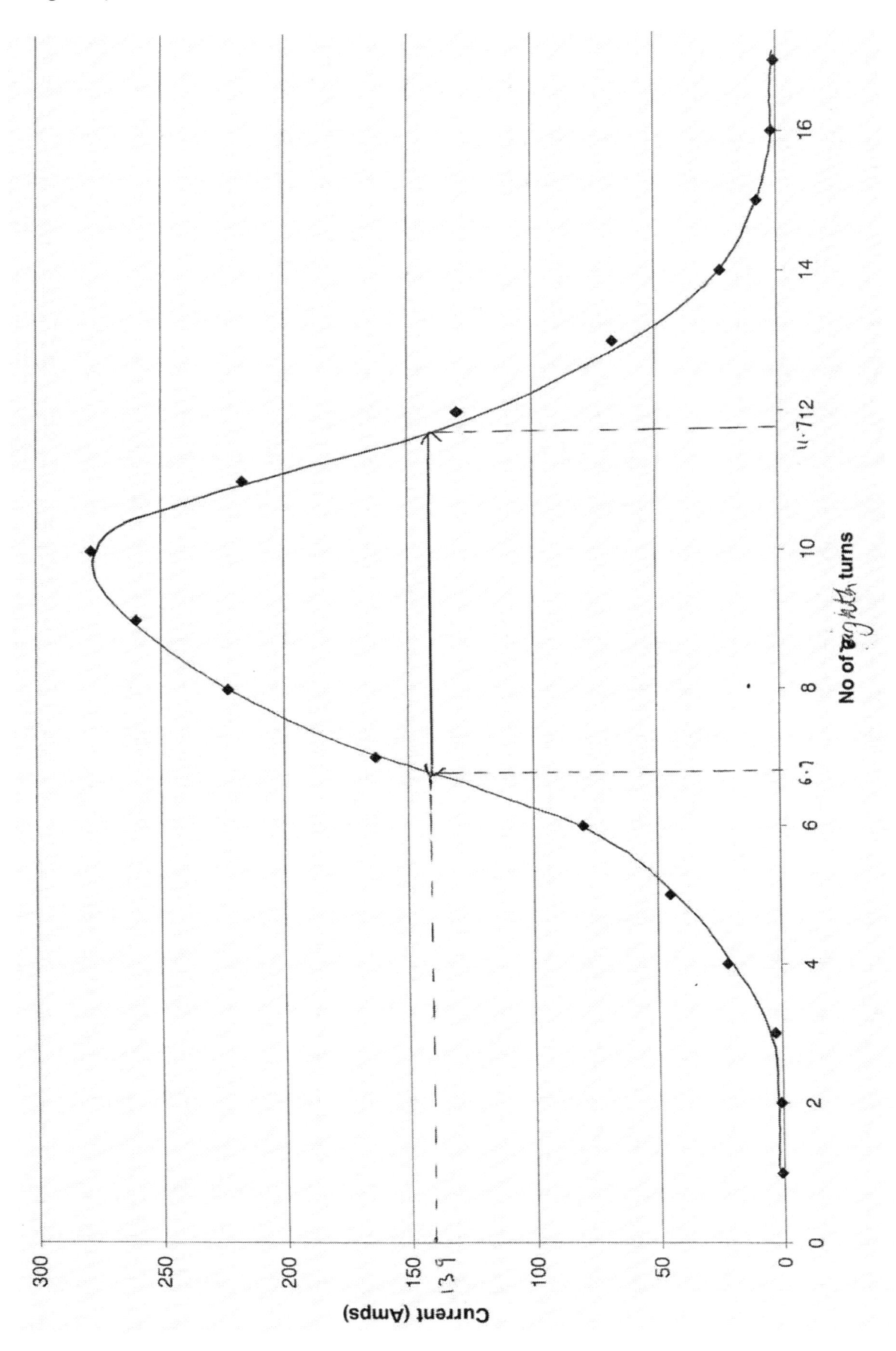

Drawing the Best-Fit Line

With the equation $F = ma$, plotting force on the y-axis and acceleration on the x-axis gives a straight line through the origin with a gradient equal to the mass. Starting from $E_k = \frac{1}{2}mv^2$, plotting the kinetic energy on the y-axis and the speed on the x-axis gives a quadratic curve passing through the origin. That isn't much use for analysis, so it's better to plot the kinetic energy on the y-axis against the speed squared on the x-axis. This graph should give a straight line of gradient equal to half the mass. Plotting combinations which give a straight line is a useful strategy since you can get useful information from the gradient. Having plotted the correct variables, how do you draw the best-fit straight line through the points?

The simplest method is to judge it by eye. Having as many points scattered above the line as below it keeps it about right. Even better is to be able to pivot the middle of the line about the correct point (using your see-through ruler). That middle point is called the centroid and is easy to calculate. It's coordinates are the mean of the x values and the mean of the y values, that is $(\bar{x}, \bar{y})$. Mark the centroid on the graph using a different symbol to distinguish it from the data points. Pivot your ruler about it and judge the best line by eye.

For the uncertainty in the gradient, again pivot the ruler about the centroid and make the gradient steeper until it looks like a poor fit. This is your steepest possible gradient, so draw it in and measure it. Repeat for the least steep gradient. This is the time-honoured way, used since the invention of graph paper (but it's rough).

However the SQA prefer to use the method shown in the graph. As before, draw in the best-fit line by eye. Line up the ruler with the best-fit line and slide it upwards and parallel, through each point, stopping at the last one. Repeat the process, but this time going downwards. This makes a parallelogram with diagonals DB and AC. Using these as the extremes of the gradient would not take the number of points into account and be an overestimate. To get the uncertainty, divide the difference by 2 (to express it as ± about the middle), then by √n (or similar) for the number of points.

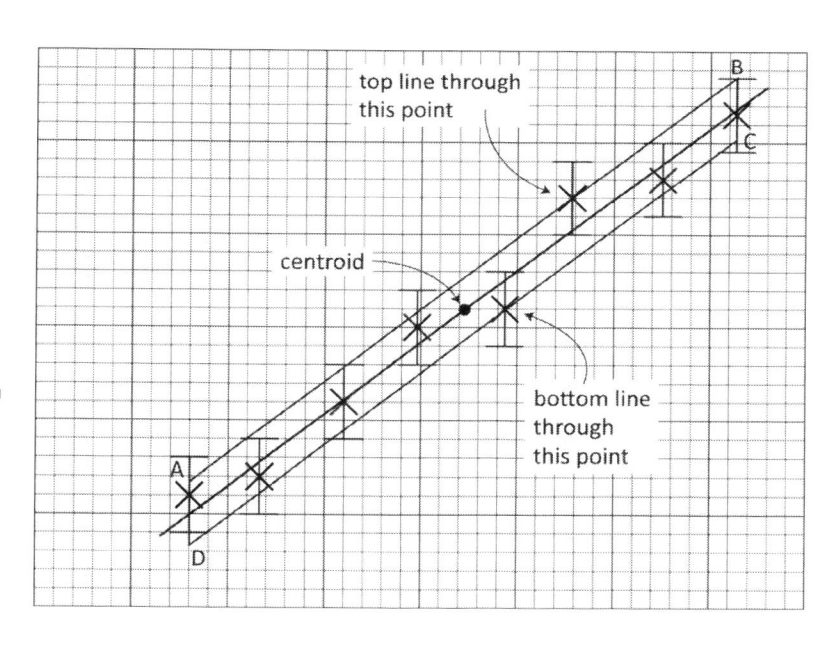

You can see why people prefer using computer software like Excel; it's less messy, there's no rubbing out, you get the best answer, it's just a few clicks of the mouse, and the SQA have no objections to its use. But suppose you weren't all that familiar with the software and it wouldn't do what you want? It can happen, so if you get frustrated, get out your pencil and draw!

What if it goes Wrong?

Here's the scenario. It's well into March, you haven't got any decent results, things just aren't working out and there's a chemistry investigation still to finish. You're expecting me to say '**don't panic**', but I'm not, because you won't buy it. Whatever you end up doing, you must submit **something**, even if it's thin and full of excuses.

It's as bad as you think and they're out to get you

Here are your options:

Option 1 Within the present investigation, research the internet for experiments similar to your own. Someone else may have had the same problems as you, and found a way through. Try to find one with lots of data so that you can construct tables, draw graphs and reach conclusions.

Option 2 Try to sort out the problems. Have you been working too much on your own? Have you kept your teacher informed of the difficulties? Do you need fresh input from someone else? Have you invested too much time on one hard experiment when there are easier ones to do? As the advert said, 'It's good to talk', and as the life coach says, 'every setback's an opportunity'.

Option 3 Abandon the investigation and do a different one; something like pendulums or LCR circuits.

Option 4 Are you the problem? Are you looking for an excuse to give up and lay the blame elsewhere? Facing up to a problem and solving it, is a character trait worth its weight in gold. This might be your chance.

At this stage, the name of the game is lots of data, and fast. If you've got data, you've got a project. Follow this with a decent write-up during the Easter holidays (grown up language, sharp conclusions and evaluation; just follow the marks allocation sheet), and your mark won't spoil the overall grade. Give it a bit of effort; believe me, it's worth it.

Printed in Great Britain
by Amazon